TRENDS AND TECHNIQUES OF GEOMORPHOLOGY

TRENDS AND TECHNIQUES OF GEOMORPHOLOGY

Bimal Dhawan

RANDOM PUBLICATIONS
NEW DELHI - 110 002 (INDIA)

Trends and Techniques of Geomorphology

ISBN 978-93-51117-43-8

Published in 2015 in India by

RANDOM PUBLICATIONS

4376-A/4B, Gali Murari Lal, Ansari Road
New Delhi-110 002
Phone: +9111-43580356, 23289044
E-mail: randomexports@gmail.com; sales@randompublications.com;
info@randompublications.com

Reprint 2021

Type Setting by: Friends Media, Delhi-110089
Digitally Printed at: Replika Press Pvt. Ltd.

Preface

Geomorphology is defined as the science of landforms with an emphasis on their origin, evolution, form, and distribution across the physical landscape. An understanding of geomorphology and its processes is therefore essential to the understanding of physical geography. Geomorphologists seek to understand why landscapes look the way they do, to understand landform history and dynamics and to predict changes through a combination of field observations, physical experiments and numerical modelling. Geomorphology is practiced within physical geography, geology, geodesy, engineering geology, archaeology and geotechnical engineering. This broad base of interest contributes to many research styles and interests within the field.

Geomorphology, studies in particular the lithosphere, and interactions with the atmosphere and hydrosphere, to understand the interconnection of various system processes. Geomorphologists are particularly interested in the potential for feedbacks between climate and tectonics mediated by geomorphic processes. Geochronology, uses dating methods to measure the rate of changes. Terrain measurement techniques, include differential GPS, remotely sensed digital terrain models and laser scanning, to quantify, study, and to generate illustrations and maps. The surface of Earth is modified by a combination of surface processes that sculpt landscapes, and geologic processes that cause tectonic uplift and subsidence, and shape the coastal geography. Surface processes comprise the action of water, wind, ice, fire, and living things on the surface of the Earth, along with chemical reactions that form soils and alter material properties, the stability and rate of change of topography under the force of gravity, and other factors, such as human alteration of the landscape. Many of these factors are strongly mediated by climate. Geologic processes include the uplift of mountain ranges, the growth of volcanoes, isostatic changes in land surface elevation, and the formation of deep sedimentary basins where the surface of Earth drops and is filled with material eroded from other parts of the landscape. The Earth surface and its topography therefore are an intersection of climatic, hydrologic, and biologic action with geologic processes. The broad-scale topographies of Earth illustrate this intersection of surface and subsurface action. Mountain belts are uplifted due to geologic processes. Denudation of these high uplifted regions produces sediment that is transported and deposited elsewhere within the landscape or off the coast. On progressively smaller scales, similar ideas apply, where individual landforms evolve in response to the balance of additive processes and subtractive processes. Often, these processes directly affect each other: ice sheets, water, and sediment are all loads that change topography through flexural isostasy. Topography can modify the local climate, for example through orographic precipitation, which in turn modifies the topography by changing the hydrologic regime in which it evolves. In addition to these broad-scale questions, geomorphologists address issues that are more specific and/or more local. Glacial geomorphologists investigate glacial deposits such as moraines, eskers, and proglacial lakes, as well as glacial erosional features, to build chronologies of both small glaciers and large ice sheets and understand their motions and effects upon the landscape. Fluvial geomorphologists focus on rivers, how they transport sediment, migrate across the landscape, cut into bedrock, respond to environmental and tectonic changes, and interact with humans. Soils geomorphologists investigate soil profiles and chemistry to learn about the history of a particular landscape and understand how climate, biota, and rock interact. Other geomorphologists study how hillslopes form and change. Still others investigate the relationships between ecology and geomorphology. Because geomorphology is defined to comprise everything related to the surface of Earth and its modification, it is a broad field with many facets. Practical applications of geomorphology include hazard assessment, river control and stream restoration, and coastal protection. Planetary geomorphology studies landforms on other terrestrial planets such as Mars. Indications of effects of wind, fluvial, glacial, mass wasting, meteor impact, tectonics and volcanic processes are studied. This effort not only helps better understand the geologic and atmospheric history of those planets but also extends geomorphological study of Earth. Planetary geomorphologists often use Earth analogues to aid in their study of surfaces of other planets.

This book is an authoritative, up-to-date review of all the major areas within geomorphology, assessing recent trends and surveying recent advances to portray the latest state of the art.

I thank all members of my team who have helped in the preparation of the book. My special thanks go to "Random Publications" who have published the book.

— *Bimal Dhawan*

Contents

Chapter 1

Geomorphological Studies of Remote Sensing

Dynamics of Geomorphology

To place geomorphology upon sound foundations for quantitative research into fundamental principles, it is proposed that geomorphic processes be treated as gravitational or molecular shear stresses acting upon elastic, plastic, or fluid earth materials to produce the characteristic varieties of strain, or failure, that constitute weathering, erosion, transportation and deposition.

Shear stresses affecting earth materials are here divided into two major categories: gravitational and molecular. Gravitational stresses activate all downslope movements of matter, hence include all mass movements, all fluvial and glacial processes. Indirect gravitational stresses activate wave-and tide-induced currents and winds. Phenomena of gravitational shear stresses are subdivided according to behavior of rock, soil, ice, water, and air as elastic or plastic solids and viscous fluids. The order of classification is generally that of decreasing internal resistance to shear and, secondarily, of laminar to turbulent flow. Molecular stresses are those induced by temperature changes, crystallization and melting, absorption and desiccation, or osmosis. These stresses act in random or unrelated directions with respect to gravity. Surficial creep results from combination of gravitational and molecular stresses on a slope. Chemical processes of solution and acid reaction are considered separately.

A fully dynamic approach requires analysis of geomorphic processes in terms of clearly defined open systems which tend to achieve steady states of operation and are self-regulatory to a large degree. Formulation of mathematical models, both by rational deduction and empirical analysis of observational data, to relate energy, mass, and time is the ultimate goal of the dynamic approach.

Geomorphology

Geomorphology is the scientific study of landforms and the processes that shape them. Geomorphologists seek to understand why landscapes look the way they do, to understand landform history and dynamics, and to predict future changes through a combination of field observations, physical experiments, and numerical modeling. Geomorphology is practiced within geography, geology, geodesy, engineering geology, archaeology, and geotechnical engineering, and this broad base of interest contributes to a wide variety of research styles and interests within the field.

Overview

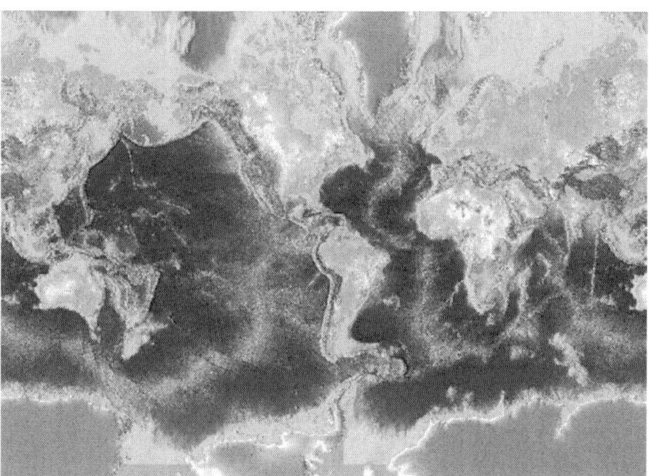

Figure 1: *Surface of the Earth*

The surface of Earth is modified by a combination of surface processes that sculpt landscapes and geologic processes that cause tectonic uplift and subsidence. Surface processes comprise the action of water, wind, ice, fire, and living things on the surface of the Earth, along with chemical reactions that form soils and alter material properties, the stability and rate of change of topography under the force of gravity, and other factors, such as (in the very recent past) human alteration of the landscape. Many of these factors are strongly mediated by climate. Geologic forcings include the uplift of mountain ranges, the growth of volcanoes, isostatic changes in land surface elevation (sometimes in response to surface processes), and the formation of deep sedimentary basins where the surface of Earth drops and is filled with material eroded from other parts of the landscape. The Earth surface and its topography therefore are an intersection of climatic, hydrologic, and biologic action with geologic processes.

The broad-scale topographies of Earth illustrate this intersection of surface and subsurface action. Mountain belts are uplifted due to geologic processes. Denudation of these high uplifted regions produces sediment that is transported and deposited elsewhere within the landscape or off the coast.

On progressively smaller scales, similar ideas apply, where individual landforms evolve in response to the balance of additive processes (uplift and deposition) and subtractive processes (subsidence and erosion). Often, these processes directly affect each other: ice sheets, water, and sediment are all loads that change topography through flexural isostasy. Topography can modify the local climate, for example through orographic precipitation, which in turn modifies the topography by changing the hydrologic regime in which it evolves. Many geomorphologists are particularly interested in the potential for feedbacks between climate and tectonics mediated by geomorphic processes.

In addition to these broad-scale questions, geomorphologists address issues that are more specific and/or more local. Glacial geomorphologists investigate glacial deposits such as moraines, eskers, and proglacial lakes, as well as glacial erosional features, to build chronologies of both small glaciers and large ice sheets and understand their motions and effects upon the landscape. Fluvial geomorphologists focus on rivers, how they transport sediment, migrate across the landscape, cut into bedrock, respond to environmental and tectonic changes, and interact with humans. Soils geomorphologists investigate soil profiles and chemistry to learn about the history of a particular landscape and understand how climate, biota, and rock interact. Other geomorphologists study how hillslopes form and change. Still others investigate the relationships between ecology and geomorphology. Because geomorphology is defined to comprise everything related to the surface of Earth and its modification, it is a broad field with many facets.

Practical applications of geomorphology include hazard assessment (such as landslide prediction and mitigation), river control and stream restoration, and coastal protection.

History

With some notable exceptions, geomorphology is a relatively young science, growing along with interest in other aspects of the Earth

Sciences in the mid 19th century. This section provides a very brief outline of some of the major figures and events in its development.

Ancient Geomorphology

Perhaps the earliest one to devise a theory of geomorphology was the polymath Chinese scientist and statesman Shen Kuo (1031-1095 AD). This was based on his observation of marine fossil shells in a geological stratum of a mountain hundreds of miles from the Pacific Ocean. Noticing bivalve shells running in a horizontal span along the cut section of a cliffside, he theorized that the cliff was once the prehistoric location of a seashore that had shifted hundreds of miles over the centuries. He inferred that the land was reshaped and formed by soil erosion of the mountains and by deposition of silt, after observing strange natural erosions of the Taihang Mountains and the Yandang Mountain near Wenzhou. Furthermore, he promoted the theory of gradual climate change over centuries of time once ancient petrified bamboos were found to be preserved underground in the dry, northern climate zone of *Yanzhou*, which is now modern day Yan'an, Shaanxi province.

Early Modern Geomorphology

The first use of the word geomorphology was likely to be in the German language when it appeared in Laumann's 1858 work. Keith Tinkler has suggested that the word came into general use in English, German and French after John Wesley Powell and W. J. McGee used it in the International Geological Conference of 1891.

An early popular geomorphic model was the *geographical cycle* or the *cycle of erosion*, developed by William Morris Davis between 1884 and 1899. The cycle was inspired by theories of uniformitarianism first formulated by James Hutton (1726–1797). Concerning valley forms, uniformitarianism depicted the cycle as a sequence in which a river cuts a valley more and more deeply, but then erosion of side valleys eventually flatten the terrain again, to a lower elevation. Tectonic uplift could start the cycle over. Many studies in geomorphology in the decades following Davis' development of his theories sought to fit their ideas into this framework for broad scale landscape evolution, and are often today termed "Davisian". Davis' ideas have largely been superseded today, mainly due to their lack of predictive power and qualitative nature, but he remains an extremely important figure in the history of the subject.

In the 1920s, Walther Penck developed an alternative model to Davis', believing that landform evolution was better described as a balance between ongoing processes of uplift and denudation, rather than

Davis' single uplift followed by decay. However, due to his relatively young death, disputes with Davis and a lack of English translation of his work his ideas were not widely recognised for many years.

These authors were both attempting to place the study of the evolution of the Earth's surface on a more generalized, globally relevant footing than had existed before. In the earlier parts of the 19th century, authors-especially in Europe-had tended to attribute the form of landscape to local climate, and in particular to the specific effects of glaciation and periglacial processes. In contrast, both Davis and Penck were seeking to emphasize the importance of evolution of landscapes through time and the generality of Earth surface processes across different landscapes under different conditions.

Quantitative Geomorphology

While Penck and Davis and their followers were writing and studying primarily in Western Europe, another, largely separate, school of geomorphology was developed in the United States in the middle years of the 20th century.

Following the early trailblazing work of Grove Karl Gilbert around the turn of the 20th century, a group of natural scientists, geologists and hydraulic engineers including Ralph Alger Bagnold, John Hack, Luna Leopold, Thomas Maddock and Arthur Strahler began to research the form of landscape elements such as rivers and hillslopes by taking systematic, direct, quantitative measurements of aspects of them and investigating the scaling of these measurements. These methods began to allow prediction of the past and future behavior of landscapes from present observations, and were later to develop into what the modern trend of a highly quantitative approach to geomorphic problems. Quantitative geomorphology can involve fluid dynamics and solid mechanics, geomorphometry, laboratory studies, field measurements, theoretical work, and full landscape evolution modeling. These approaches are used to understand weathering and the formation of soils, sediment transport, landscape change, and the interactions between climate, tectonics, erosion, and deposition.

Contemporary Geomorphology

Today, the field of geomorphology encompasses a very wide range of different approaches and interests. Modern researchers aim to draw out quantitative "laws" that govern Earth surface processes, but equally, recognize the uniqueness of each landscape and environment in which these processes operate. Particularly important realizations in contemporary geomorphology include:

1) that not all landscapes can be considered as either "stable" or "perturbed", where this perturbed state is a temporary displacement away from some ideal target form. Instead, dynamic changes of the landscape are now seen as an essential part of their nature.
2) that many geomorphic systems are best understood in terms of the stochasticity of the processes occurring in them, that is, the probability distributions of event magnitudes and return times. This in turn has indicated the importance of chaotic determinism to landscapes, and that landscape properties are best considered statistically. The same processes in the same landscapes does not always lead to the same end results.

Processes

Modern geomorphology focuses on the quantitative analysis of interconnected processes. Modern advances in geochronology, in particular cosmogenic radionuclide dating, optically stimulated luminescence dating and low-temperature thermochronology have enabled us for the first time to measure the rates at which geomorphic processes occur. At the same time, the use of more precise physical measurement techniques, including differential GPS, remotely sensed digital terrain models and laser scanning techniques, have allowed quantification and study of these processes as they happen. Computer simulation and modeling may then be used to test our understanding of how these processes work together and through time.

Geomorphically relevant processes generally fall into (1) the production of regolith by weathering and erosion, the transport of that material, and its eventual deposition. Although there is a general movement of material from uplands to lowlands, erosion, transport, and deposition often occur in closely-spaced tandem all across the landscape. The nature of the processes investigated by geomorphologists is strongly dependent on the landscape or landform under investigation and the time and length scales of interest.

However, the following non-exhaustive list provides a flavor of the landscape elements associated with some of these.

Primary surface processes responsible for most topographic features include wind, waves, chemical dissolution, mass wasting, groundwater movement, surface water flow, glacial action, tectonism, and volcanism. Other more exotic geomorphic processes might include periglacial (freeze-thaw) processes, salt-mediated action, or extraterrestrial impact.

Fluvial Processes

Rivers and streams are not only conduits of water, but also of sediment. The water, as it flows over the channel bed, is able to mobilize sediment and transport it downstream, either as bed load, suspended load or dissolved load. The rate of sediment transport depends on the availability of sediment itself and on the river's discharge.

Rivers are also capable of eroding into rock and creating new sediment, both from their own beds and also by coupling to the surrounding hillslopes. In this way, rivers are thought of as setting the base level for large scale landscape evolution in nonglacial environments. Rivers are key links in the connectivity of different landscape elements.

As rivers flow across the landscape, they generally increase in size, merging with other rivers. The network of rivers thus formed is a drainage system and is often dendritic, but may adopt other patterns depending on the regional topography and underlying geology.

Aeolian Processes

Aeolian processes pertain to the activity of the winds and more specifically, to the winds' ability to shape the surface of the Earth. Winds may erode, transport, and deposit materials, and are effective agents in regions with sparse vegetation and a large supply of unconsolidated sediments. Although water and mass flow tend to mobilize more material than wind in most environments, aeolian processes are important in arid environments such as deserts.

Hillslope Processes

Soil, regolith, and rock move downslope under the force of gravity via creep, slides, flows, topples, and falls. Such mass wasting occurs on both terrestrial and submarine slopes, and has been observed on Earth, Mars, Venus, Titan and Iapetus. Ongoing hillslope processes can change the topology of the hillslope surface, which in turn can change the rates of those processes. Hillslopes that steepen up to certain critical thresholds are capable of shedding extremely large volumes of material very quickly, making hillslope processes an extremely important element of landscapes in tectonically active areas. On Earth, biological processes such as burrowing or tree throw may play important roles in setting the rates of some hillslope processes.

Glacial Processes

Glaciers, while geographically restricted, are effective agents of landscape change. The gradual movement of ice down a valley causes abrasion and plucking of the underlying rock. Abrasion produces fine sediment, termed glacial flour. The debris transported by the glacier, when the glacier recedes, is termed a moraine. Glacial erosion is responsible for U-shaped valleys, as opposed to the V-shaped valleys of fluvial origin.

The way glacial processes interact with other landscape elements, particularly hillslope and fluvial processes, is an important aspect of Plio-Pleistocene landscape evolution and its sedimentary record in many high mountain environments. Environments that have been relatively recently glaciated but are no longer may still show elevated landscape change rates compared to those that have never been glaciated. Nonglacial geomorphic processes which nevertheless have been conditioned by past glaciation are termed paraglacial processes. This concept contrasts with periglacial processes, which are directly driven by formation or melting of ice or frost.

Tectonic Processes

Tectonic effects on geomorphology can range from scales of millions of years to minutes or less. The effects of tectonics on landscape are heavily dependent on the nature of the underlying bedrock fabric that more less controls what kind of local morphology tectonics can shape. Earthquakes can, in terms of minutes, submerge large extensions creating new wetlands. Isostatic rebound can account for significant changes over thousand or hundreds of years, and allows erosion of a mountain belt to promote further erosion as mass is removed from the chain and the belt uplifts. Long-term plate tectonic dynamics give rise to orogenic belts, large mountain chains with typical lifetimes of many tens of millions of years, which form focal points for high rates of fluvial and hillslope processes and thus long-term sediment production.

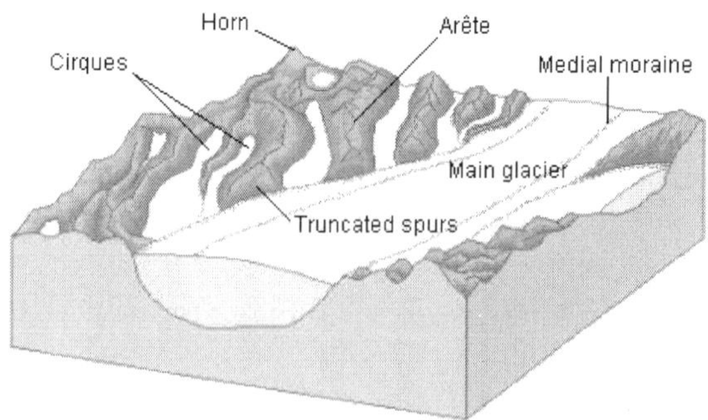

***Figure 2:** Features of a glacial landscape*

Features of deeper mantle dynamics such as plumes and delamination of the lower lithosphere have also been hypothesised to play important roles in the long term (> million year), large scale (thousands of km) evolution of the Earth's topography. Both can promote surface uplift through isostasy as hotter, less dense, mantle rocks displace cooler, denser, mantle rocks at depth in the Earth.

Igneous Processes

Figure 3: *Some landscapes are dominated by igneous processes like Villarrica National Park in the picture.*

Both volcanic (eruptive) and plutonic (intrusive) igneous processes can have important impacts on geomorphology. The action of volcanoes tends to rejuvenize landscapes, covering the old land surface with lava and tephra, releasing pyroclastic material and forcing rivers through new paths. The cones built by eruptions also build substantial new topography, which can be acted upon by other surface processes.

Biological Processes

The interaction of living organisms with landforms, or biogeomorphologic processes, can be of many different forms, and is probably of profound importance for the terrestrial geomorphic system as a whole. Biology can influence very many geomorphic processes, ranging from biogeochemical processes controlling chemical weathering, to the influence of mechanical processes like burrowing and tree throw on soil development, to even controlling global erosion rates through modulation of climate through carbon dioxide balance. Terrestrial landscapes in which the role of biology in mediating surface processes can be definitively excluded are extremely rare, but may hold important information for understanding the geomorphology of other planets, such as Mars.

Scales in Geomorphology

Different geomorphological processes dominate at different spatial and temporal scales. Moreover, scales on which processes occur may determine the reactivity or otherwise of landscapes to changes in driving forces such as climate or tectonics. These ideas are key to the study of geomorphology today.

Figure 4: *Beaver dams, as this one in Tierra del Fuego, constitute a specific form of zoogeomorphology, a type of biogeomorphology*

To help categorize landscape scales some geomorphologists might use the following taxonomy:

- 1st-Continent, ocean basin, climatic zone (~10,000,000 km^2)
- 2nd-Shield, e.g. Baltic Shield, or mountain range (~1,000,000 km^2)
- 3rd-Isolated sea, Sahel (~100,000 km^2)
- 4th-Massif, e.g. Massif Central or Group of related landforms, e.g., Weald (~10,000 km^2)
- 5th-River valley, Cotswolds (~1,000 km^2)
- 6th-Individual mountain or volcano, small valleys (~100 km^2)
- 7th-Hillslopes, stream channels, estuary (~10 km^2)
- 8th-gully, barchannel (~1 km^2)
- 9th-Meter-sized features.

Overlap with other Fields

There is a considerable overlap between geomorphology and other fields. Deposition of material is extremely important in sedimentology. Weathering is the chemical and physical disruption of earth materials in place on exposure to atmospheric or near surface agents, and is typically studied by soil scientists and environmental chemists, but is an essential component of geomorphology because it is what provides the material that can be moved in the first place. Civil and environmental engineers are concerned with erosion and sediment transport, especially related to canals, slope stability (and natural hazards), water quality, coastal environmental management, transport of contaminants, and stream restoration. Glaciers can cause extensive erosion and deposition in a short period of time, making them extremely important entities in the high latitudes and meaning that they set the conditions in the headwaters of mountain-born streams; glaciology therefore is important in geomorphology.

Geomorphic Processes

Concept

The surface of Earth is covered with various landforms, a number of which are discussed in various entries throughout this book. This essay is devoted to the study of landforms themselves, a subdiscipline of the geologic sciences known as geomorphology. The latter, as it has evolved since the end of the nineteenth century, has become an interdisciplinary study that draws on areas as diverse as plate tectonics, ecology, and meteorology. Geomorphology is concerned with the shaping of landforms, through such processes as subsidence and uplift, and with the classification and study of such landforms as mountains, volcanoes, and islands.

An Evolving Area of Study: Geomorphology is an area of geology concerned with the study of landforms, with the forces and processes that have shaped them, and with the description and classification of various physical features on Earth. The term, which comes from the Greek words *geo,* or "Earth," and *morph,* meaning "form," was coined in 1893 by the American geologist William Morris Davis (1850-1934), who is considered the father of geomorphology.

During Davis's time, geomorphology was concerned primarily with classifying different structures on Earth's surface, examples of which include mountains and islands, discussed later in this essay. This view of geomorphology as an essentially descriptive, past-oriented area of study closely aligned with historical geology prevailed throughout the late nineteenth and early twentieth centuries.

By the mid-twentieth century, however, the concept of geomorphology inherited from Davis had fallen into disfavor, to be replaced by a paradigm, or model, oriented toward physical rather than historical geology. (These two principal branches of geology are concerned, in the first instance, with Earth's past and the processes that shaped it and, in the second instance, with Earth's current physical features and the processes that continue to shape it.)

Rethinking Geomorphology: As reconceived in the 1950s and thereafter, geomorphology became an increasingly exact science. As has been typical of many sciences in their infancy, early geomorphology focused on description rather than prediction and tended to approach its subject matter in a qualitative fashion. The term qualitative suggests a comparison between qualities that are not defined precisely, such as "fast" and "slow" or "warm" and "cold." On the other hand, a quantitative approach, as has been implemented for geomorphology from the mid-twentieth century onward, centers on a comparison between precise quantities—for instance, 10 lb. (4.5 kg) versus 100 lb. (45 kg) or 50 MPH (80.5 km/h) versus 120 MPH (193 km/h).

As part of its shift in focus, geomorphology began to treat Earth's physical features as systems made up of complex and ongoing interactions. This view fell into line with a general emphasis on the systems concept in the study of Earth. As geomorphology evolved, it became more interdisciplinary, as we shall see. This, too, was part of an overall trend in the earth sciences toward an approach that viewed subjects in broad, cross-disciplinary terms as opposed to a narrow focus on specific areas of study.

Landforms and Processes: Two concerns are foremost within the realm of geomorphology, and these concerns reflect the stages of its history. First, in line with Davis's original conception of geomorphology as an area of science devoted to classifying and describing natural features, there is its concern with topography. The latter may be defined as the configuration of Earth's surface, including its relief (elevation and other in equalities) as well as the position of physical features.

These physical features are called landforms, examples of which include mountains, plateaus, and valleys. Geomorphology always has involved classification, and early scientists working in this subdiscipline addressed the classification of landforms. Other systems of classification, however, are not so concerned with cataloging topographical features themselves as with differentiating the processes that shaped them. This brings us to the other area of interest in geomorphology: the study of how landforms came into being.

Shaping the Earth: Among the processes that drive the shaping of landforms is plate tectonics, or the shifting of large, movable segments of lithosphere (the crust and upper layer of Earth's mantle). Plate tectonics is discussed in detail within its own essay and more briefly in other areas throughout this book, as befits its status as one of the key areas of study in the earth sciences.

Other processes also shape landforms. Included among these processes are weathering, the breakdown of rocks and minerals at or near the surface of Earth due to physical or chemical processes; erosion, the movement of soil and rock due to forces produced by water, wind, glaciers, gravity, and other influences; and mass wasting or mass movement, the transfer of earth material, by processes that include flow, slide, fall, and creep, down slopes. Also of interest are fluvial and eolian processes (those that result from water flow and wind, respectively) as well as others related to glaciers and coastal formations.

Human activity also can play a significant role in shaping Earth. This effect may be direct, as when the construction of cities, the building of dams, or the excavation of mines alters the landscape. On the other hand, it can be indirect. In the latter instance, human activity in the biosphere exerts an impact, as when the clearing of forest land or the misuse of crop land results in the formation of a dust bowl.

Interdisciplinary Studies: As noted earlier, geomorphology is characteristic of the earth sciences as a whole in its emphasis on an interdisciplinary approach. As is true of earth scientists in general, those studying landforms and the processes that shape them do not work simply in one specialty. Among the areas of interest in geomorphology are, for example, deep-sea geomorphology, which draws on oceanography, and planetary geomorphology, the study of landscapes on other planets.

When studying coastal geomorphology, a geologist may draw on realms as diverse as fluid mechanics (an area of physics that studies the behavior of gases and liquids at rest and in motion) and sedimentology. The investigation of such processes as erosion and mass wasting calls on knowledge in the atmospheric sciences as well as the physics and chemistry of soil. It is almost inevitable that a geomorphologic researcher will draw on geophysics as well as on such subspecialties as volcanology. These studies may go beyond the "hard sciences," bringing in such social sciences as geography.

Real-Life Applications

Subsidence: Subsidence refers to the process of subsiding (settling or descending), on the part of either an air column or the solid earth, or, in the case of solid earth, to the resulting formation or depression. Subsidence in the atmosphere is discussed briefly in the entry Convection. Subsidence that occurs in the solid earth, known as geologic subsidence, is the settling or sinking by a body of rock or sediment. (The latter can be defined as material deposited at or near Earth's surface from a number of sources, most notably preexisting rock.)

As noted earlier, many geomorphologic processes can be caused either by nature or by human beings. An example of natural subsidence takes place in the aftermath of an earthquake, during which large areas of solid earth may simply drop by several feet. Another example can be observed at the top of a volcano some time after it has erupted, when it has expelled much of its material (i.e., magma) and, as a result, has collapsed. Natural subsidence also may result from cave formation in places where underground water has worn away limestone. If the water erodes too much limestone, the ceiling of the cave will subside, usually forming a sinkhole at the surface. The sinkhole may fill with water, making a lake; the formation of such sinkholes in many spots throughout an area (whether the sinkholes become lakes or not), is known as karst topography.

In places where the bedrock is limestone—particularly in the sedimentary basins of rivers—karst topography is likely to develop. The United States contains the most extensive karst region in the world, including the Mammoth cave system in Kentucky. Karst topography is very pronounced in the hills of southern China, and karst landscapes have been a prominent feature of Chinese art for centuries. Other extensive karst regions can be found in southern France, Central America, Turkey, Ireland, and England.

Man-Made Subsidence: Man-made subsidence often ensues from the removal of groundwater or fossil fuels, such as petroleum or coal. Groundwater removal can be perfectly safe, assuming the area experiences sufficient rainfall to replace, or recharge, the lost water. If recharging does not occur in the necessary proportions, however, the result will be the eventual collapse of the aquifer, a layer of rock that holds groundwater.

In so-called room-and-pillar coal mining, pillars, or vertical columns, of coal are left standing, while the areas around them are extracted. This method maintains the ceiling of the "room" that has been mined of its coal. After the mine is abandoned, however, the pillar eventually may experience so much stress that it breaks, leading to the collapse of the mined room. As when the ceiling of a cave collapses, the subsidence of a coal mine leaves a visible depression above ground.

Uplift: As its name implies, uplift describes a process and results opposite to those of subsidence. In uplift the surface of Earth rises, owing either to a decrease in downward force or to an increase in upward force. One of the most prominent examples of uplift is seen when plates collide, as when India careened into the southern edge of the Eurasian landmass some 55 million years ago. The result has been a string of mountain ranges, including the Himalayas, Karakoram Range, and Hindu Kush, that contain most of the world's tallest peaks.

Plates move at exceedingly slow speeds, but their mass is enormous. This means that their inertia (the tendency of a moving object to keep moving unless acted upon by an outside force) is likewise gargantuan in scale. Therefore, when plates collide, though they are moving at a rate equal to only a few inches a year, they will keep pushing into each other like two automobiles crumpling in a head-on collision. Whereas a car crash is over in a matter of seconds, however, the crumpling of continental masses takes place over hundreds of thousands of years.

When sea floor collides with sea floor, one of the plates likely will be pushed under by the other one, and, likewise, when sea floor collides with continental crust, the latter will push the sea floor under. This results in the formation of volcanic mountains, such as the Andes of South America or the Cascades of the Pacific Northwest, or volcanic islands, such as those of Japan, Indonesia, or Alaska's Aleutian chain.

Isostatic Compensation: In many other instances, collision, compression, and extension cause uplift. On the other hand, as noted, uplift may result from the removal of a weight. This occurs at the end of an ice age, when glaciers as thick as 1.9 mi. (3 km) melt, gradually removing a vast weight pressing down on the surface below.

This movement leads to what is called isostatic compensation, or isostatic rebound, as the crust pushes upward like a seat cushion rising after a person is longer sitting on it. Scandinavia is still experiencing uplift at a rate of about 0.5 in. (1 cm) per year as the after-effect of glacial melting from the last ice age. The latter ended some 10,000 years ago, but in geologic terms this is equivalent to a few minutes' time on the human scale.

Islands: Geomorphology, as noted earlier, is concerned with landforms, such as mountains and volcanoes as well as larger ones, including islands and even continents. Islands present a particularly interesting area of geomorphologic study. In general, islands have certain specific characteristics in terms of their land structure and can be analyzed from the standpoint of the geosphere, but particular islands also have unique ecosystems, requiring an interdisciplinary study that draws on botany, zoology, and other subjects.

In addition, there is something about an island that has always appealed to the human imagination, as evidenced by the many myths, legends, and stories about islands. Some examples include Homer's *Odyssey,* in which the hero Odysseus visits various islands in his long wanderings; Thomas More's *Utopia,* describing an idealized island republic; *Robinson Crusoe,* by Daniel Defoe, in which the eponymous hero lives for many years on an island with no companion but the trusty native Friday; *Treasure Island,* by Robert Louis Stevenson, in which the island is the focus of a treasure hunt; and Mark Twain's *Adventures of Huckleberry Finn,* depicting Jackson Island in the Mississippi River, to which Huckleberry Finn flees to escape "civilization."

One of the favorite subjects of cartoonists is that of a castaway stranded on a desert island, a mound of sand with no more than a single tree. Movies, too, have long portrayed scenarios, from the idyllic to the brutal, that take place on islands, particularly deserted ones, a notable example being *Cast Away* (2000). A famous line by the English poet John Donne (1572-1631) warns that "no man is an island," implying that many wish they could enjoy the independence suggested by the concept of an island. Within the Earth system, however, nothing is fully independent, and, as we shall see, this is certainly the case where islands are concerned.

The Islands of Earth: Earth has literally tens of thousands of islands. Just two archipelagos (island chains), those that make up the Philippines and Indonesia, include thousands of islands each. While there are just a few dozen notable islands on Earth, many more dot the planet's seas and oceans. The largest are these:

- Greenland (Danish, northern Atlantic): 839,999 sq. mi.(2,175,597 sq km)
- New Guinea (divided between Indonesia and Papua New Guinea, western Pacific): 316,615 sq. mi. (820,033 sq km)
- Borneo (divided between Indonesia and Malaysia, western Pacific): 286,914 sq. mi. (743,107 sq km)
- Madagascar (Malagasy Republic, western Indian Ocean): 226,657 sq. mi. (587,042 sq km)
- Baffin (Canadian, northern Atlantic): 183,810 sq. mi. (476,068 sq km)
- Sumatra (Indonesian, northeastern Indian Ocean): 182,859 sq. mi. (473,605 sq km).

The list could go on and on, but it stops at Sumatra because the next-largest island, Honshu (part of Japan), is less than half as large, at 88,925 sq. mi. (230,316 sq km). Clearly, not all islands are created equal, and though some are heavily populated or enjoy the status of independent nations (e.g., Great Britain at number eight or Cuba at number 15), they are not necessarily the largest. On the other hand, some of the largest are among the most sparsely populated.

Of the 32 largest islands in the world, more than a third are in the icy northern Atlantic and Arctic, with populations that are small or practically nonexistent. Greenland's population, for instance, was just over 59,000 in 1998, while that of Baffin Island was about 13,200. On both islands, then, each person has about 14 frozen sq. mi. (22 sq km) to himself or herself, making them among the most sparsely populated places on Earth.

Continents, Oceans, and Islands: Australia, of course, is not an island but a continent, a difference that is not related directly to size. If Australia *were* an island, it would be by far the largest. Australia is regarded as a continent, however, because it is one of the principal landmasses of the Indo-Australian plate, which is among a handful of major continental plates on Earth. Whereas continents are more or less permanent (though they have experienced considerable rearrangement over the eons), islands come and go, seldom lasting more than 10 million years. Erosion or rising sea levels remove islands, while volcanic explosions can create new ones, as when an eruption off the coast of Iceland resulted in the formation of an island, Surtsey, in 1963.

Islands are of two types, continental and oceanic. Continental islands are part of continental shelves (the submerged, sloping ledges of continents) and may be formed in one of two ways. Rising ocean waters either cover a coastal region, leaving only the tallest mountains exposed as islands or cut off part of a peninsula, which then becomes an island. Most of Earth's significant islands are continental and are easily spotted as such, because they lie at close proximity to continental landmasses. Many other continental islands are very small, however; examples include the barrier islands that line the East Coast of the United States. Formed from mainland sand brought to the coast by rivers, these are technically not continental islands, but they more clearly fit into that category than into the grouping of oceanic islands.

Oceanic islands, of which the Hawaiian-Emperor island chain and the Aleutians off the Alaskan coast are examples, form as a result of volcanic activity on the ocean floor. In most cases, there is a region of high volcanic activity, called a hot spot, beneath the plates, which move across the hot spot. This is the situation in Hawaii, and it explains why the volcanoes on the southern islands are still active while those to the north are not: the islands themselves are moving north across the hot spot. If two plates converge and one subducts, a deep trench with a parallel chain of volcanic islands may develop. Exemplified by the Aleutians, these chains are called island arcs.

Island Ecosystems: The ecosystem, or community of all living organisms, on islands can be unique owing to their separation from continents. The number of life-forms on an island is relatively small and can encompass some unusual circumstances compared with the larger ecosystems of continents.

Ireland, for instance, has no native snakes, a fact "explained" by the legend that Saint Patrick drove them away. Hawaii and Iceland are also blessedly free of serpents.

Oceanic islands, of course, tend to have more unique ecosystems than do continental islands. The number of land-based animal life-forms is necessarily small, whereas the varieties of birds, flying insects, and surrounding marine life will be greater owing to those creatures' mobility across water. Vegetation is relatively varied, given the fact that winds, water currents, and birds may carry seeds. Nonetheless, ecosystems of islands tend to be fairly delicate and can be upset by the human introduction of new predators (e.g., dogs) or new creatures to consume plant life (e.g., sheep). These changes sometimes can have disastrous effects on the overall balance of life on islands. Overgrazing may even open up the possibility of erosion, which has the potential of bringing an end to an island's life.

Resulting Landforms of Geomorphology

Landform

A landform or physical feature in the earth sciences and geology sub-fields, comprises a geomorphological unit, and is largely defined by its surface form and location in the landscape, as part of the terrain, and as such, is typically an element of topography. Landform elements also include seascape and oceanic waterbody interface features such as bays, peninsulas, seas and so forth, including sub-aqueous terrain features such as submersed mountain ranges, volcanoes, and the great ocean basins.

Physical Characteristics

Landforms are categorised by characteristic physical attributes such as elevation, slope, orientation, stratification, rock exposure, and soil type. Gross *physical features or landforms* include intuitive elements such as berms, mounds, hills, ridges, cliffs, valleys, rivers, peninsulas and numerous other structural and size-scaled (i.e. ponds vs. lakes, hills vs. mountains) elements including various kinds of inland and oceanic waterbodies and sub-surface features.

Hierarchy of Classes

Oceans and continents exemplify the highest-order landforms. Landform elements are parts of a high-order landforms that can be further identified and systematically given a cohesive definition such as hill-tops, shoulders, saddles, foreslopes and backslopes.

Some generic landform elements including: pits, peaks, channels, ridges, passes, pools and plains, may be extracted from a digital elevation model using some automated techniques where the data has been gathered by modern satellites and stereoscopic aerial surveillance cameras. Until recently, compiling the data found in such data sets required time consuming and expensive techniques of many man-hours.

Terrain (or *relief*) is the third or vertical dimension of *land surface*. Topography is the study of terrain, although the word is often used as a synonym for relief itself. When relief is described underwater, the term bathymetry is used. In cartography, many different techniques are used to describe relief, including contour lines and TIN (Triangulated irregular network).

Elementary landforms (segments, facets, relief units) are the smallest homogeneous divisions of the land surface, at the given scale/ resolution. These are areas with relatively homogenous morphometric properties, bounded by lines of discontinuity. A plateau or a hill can be observed at various scales ranging from few hundred meters to hundreds of kilometers. Hence, the spatial distribution of landforms is often scale-dependent as is the case for soils and geological strata.

A number of factors, ranging from plate tectonics to erosion and deposition, can generate and affect landforms. Biological factors can also influence landforms— for example, note the role of vegetation in the development of dune systems and salt marshes, and the work of corals and algae in the formation of coral reefs. Landforms do not include man-made features, such as canals, ports and many harbors; and geographic features, such as deserts, forests, grasslands, and impact craters.

Many of the terms are not restricted to refer to features of the planet Earth, and can be used to describe surface features of other planets and similar objects in the Universe. Examples are mountains, polar caps, and valleys, which are found on all of the terrestrial planets.

Landform Construction and Destruction

Megageomorphology: In keeping with the regional scale of our examination of major geomorphic provinces we will begin this course by considering tectonic landscapes on the scale of whole mountain ranges and some of the general processes that shape them.

Geomorphologists have also traditionally divided landforms into constructional and destructional categories.

Constructional landforms are those that have been or are being built (increasing in mass, height, or area). Examples include tectonic forms such as fault scarps, fault block mountains and valleys (horsts and grabens), rift valleys, monoclines, salt domes, and a variety of volcano landforms.

In general, constructional landforms are more common in regions of active tectonism, usually associated with plate margins. Destructional landforms are those that are decreasing in mass, height, or area over time and are primarily the products of erosion and weathering. In truth, all landforms are in part destructional because erosion affects all landforms that lie above sea level on the Earth's surface. However, many landforms are primarily the result of such processes.

More recently, geomorphologists have come to view all landscapes as the products of some combination of three factors:

- tectonics (orogeny, epeirogeny, extension, volcanism)
- erosion
- climate.

None of these factors operates independently-all are in some way functions of the others:

- Tectonics are clearly affected by erosion-removal of crustal mass by erosion leads to isostatic rebound and uplift. In general, about 80% of the height of a landmass removed by erosion will be replaced from below by isostatic readjustment. Some geologists refer to the "pull of erosion" to describe this phenomenon.
- Uplift accelerates erosion, which, as a general rule, increases in intensity with increasing elevation. High elevations also host less vegetation, which armors the land against erosion at low elevations.
- Erosion is a function of climate in many ways. Humid climates as a rule erode faster than arid climates and the effects of erosion on the landscape differ. Humid climates tend to result in the formation of deeply incised valleys. Arid regions erode more evenly. Cold climates result in mountain glaciers that can carve extremely steep topographic features. In fact, mountain glaciers may be Earth's most effective agents of erosion. Extremely frigid climates host large ice sheets that do very little erosion because they are frozen to the bedrock on which they rest.
- Climate itself is greatly affected by tectonism. High landmasses change the balance of heating on the surface and block or redirect the flow of weather systems. As a landmass is elevated

its climate cools and may become more humid. This in turn can accelerate erosion. Windward of a mountain range the climate is usually humid as ascending air masses cool and drop their moisture. Leeward of the range is a rain shadow and very arid conditions.

Isostatic Mountain Building: Climate and erosion can also work together to create uplifted mountain peaks. An arid plateau will support few rivers and tend to erode slowly and uniformly. As material is removed across the plateau isostatic readjustment will cause some uplift and the average land surface will slowly be lowered. If however, the plateau is in moist climate with abundant rivers or glaciers, then abundant, deep valleys can form, removing a large mass of crust, but leaving some crust standing as peaks. Isostatic readjustment will elevate the plateau so that its average mass will be lower than before erosion, but this can result in the peaks being elevated to a higher level than before.

The Three Stages of Mountains: It is probably incorrect to say the mountains are uplifted and then eroded back to sea level. Rather, isostacy, climate and erosion interact to create three stages in the life of a mountain range.

1. Formative Stage-a tectonic event thickens or heats the crust and creates uplift with rates that exceed erosion rates. As the topography rises erosion rates increase.
2. Steady State-erosion rates rise to match uplift rates or uplift slows to match erosion. A steady state results with uplift and erosion rates balanced. Because uplift rates and erosion rates create a negative feedback loop (increasing uplift raises the mountains which intensifies erosion and lowers the mountains) the equilibrium elevation of the mountains may remain stable for millions of years.
3. Decline-if uplift diminishes sufficiently erosion may come to dominate and slowly lower the elevation of the range. This is a slow process because isostatic rebound will replace much of the elevation lost to erosion. By the time a mountain range is worn back to the average level of continental crust, a thickness of crust several times greater than the height of the mountains at their highest can be removed by erosion. For example, it is estimated from mineral studies that the metamorphic rocks of New York City and Connecticut were once buried under about 7 miles of overlying rock. This is not to say that the mountains above them were 7 miles high, only that this thickness of crust

has been pulled upward by isostatic readjustment over the last 450 million years.

Major Geomorpho-tectonic Regions

- Coastal Plains-these are extensive regions of flat-lying to gently seaward dipping sediments that build up as a passive continental margin subsides over millions of years. They are areas of very low relief. Generally, coastal plains are absent in tectonically active areas because active uplift of the coastline prohibits their development.
- Orogenic Belts-these are elongate regions that form from collisional tectonics along active continental margins. In presently active orogenic belts the metamorphic rocks may not yet be exposed on the surface-only after extensive erosion of the overlying, intensely deformed sedimentary cover are the metamorphic roots of the mountains exposed. Where they are generated by subduction of oceanic crust beneath the continent they are highly intruded by magmatic plutons.
- Fold and Thrust Belts-marginal to orogenic belts are regions where the sedimentary cover has been folded and thrust faulted by compressional stresses without metamorphism. Low angle thrust faults create blocks of sedimentary crust stacked one on another, suggesting large amounts of shortening of the upper crust (often refered to as "thin-skined tectonics"). Erosion of ancient fold and thrust belts near sea level produces a distinctive ridge and valley topography with surface expressions of underlying syncline and anticline folds.
- Plateaus-these are tectonically elevated regions of undeformed sedimentary rock (the Tibetan Plateau and the Chilean Altiplano are a somewhat different type of plateau-elevated regions within an orogenic belt created by slow erosion due to rain shadow climates). Plateaus developed in in humid regions can be very mountainous due to stream dissection. An extreme form of stream dissection is seen in the canyonlands of the Colourado Plateau, where rivers such as the Colourado have carved mile-deep canyons such as the Grand Canyon.
- Stable Interior/Shield-the interior regions of continents are generally low relief areas with very mild stream dissection. Shield areas in the center of a continent are underlain by ancient, deeply eroded metamorphic and metaigneous rocks. Surrounding the shield regions are areas of the stable interior covered with a thin veneer of nearly horizontal sedimentary strata.

These strata are very subtly deformed into broad structurally upwarped regions called domes and downwarped regions called basins which generally do not have any topographic expression.

- Extensional Regions-regions of high mantle heat flow beneath continental crust can cause broad crustal doming and the formation of fault block mountains bounded by grabens and half grabens. Large grabens are also known as rift valleys when they are associated with continental rifting. These features form as the crust is thinned and stretched, creating tensional forces and normal faults. The extensive Basin and Range region of Utah and Nevada in the western US has been generated in the last 17 million years by extension of the western US by as much as 100 miles in the east-west direction. Extensional tectonics are often accompanied by igneous intrusives and surface eruptions as magma works its way upward along faults in the thinned crust.

Morphometric Analysis

Morphometric analysis, quantitative description and analysis of landforms as practiced in geomorphology that may be applied to a particular kind of landform or to drainage basins and large regions generally.

Formulas for right circular cones have been fitted to the configurations of alluvial fans, logarithmic spirals have been used to describe certain shapes of beaches, and drumlins, spoon-shaped glacial landforms, have been found to accord to the form of the lemniscate curve. With regard to drainage basins, many quantitative measures have been developed to describe valley side and channel slopes, relief, area, drainage network type and extent, and other variables. Attempts to correlate statistically parameters defining drainage basin characteristics and basin hydrology, as in studies of sediment yield, are generally designated as morphometric analyses.

Morphometrics

Morphometrics Morphometrics refers to the quantitative analysis of *form*, a concept that encompasses size and shape. Morphometric analyses are commonly performed on organisms, and are useful in analyzing their fossil record, the impact of mutations on shape, developmental changes in form, covariances between ecological factors and shape, as well for estimating quantitative-genetic parameters of shape.

Morphometrics can be used to quantify a trait of evolutionary significance, and by detecting changes in the shape, deduce something of their ontogeny, function or evolutionary relationships. A major objective of morphometrics is to statistically test hypotheses about the factors that affect shape.

"Morphometrics", in the broader sense of the term, is also used to precisely locate certain areas of featureless organs such as the brain, and is used in describing the shapes of other things.

Forms of Morphometrics

Three general approaches to form are usually distinguished: traditional morphometrics, landmark-based morphometrics and outline-based morphometrics.

"Traditional" Morphometrics

Traditional morphometrics analyzes lengths, widths, masses, angles, ratios and areas (image). In general, traditional morphometric data are measurements of size. A drawback of using many measurements of size is that most will be highly correlated; as a result, there are few independent variables despite the many measurements. For instance, tibia length will vary femur length and also with humerus and ulna length and even with measurements of the head. Traditional morphometric data are nonetheless useful when either absolute or relative sizes are of particular interest, such as in studies of growth. These data are also useful when size measurements are of theoretical importance such as body mass and limb cross-sectional area and length in studies of functional morphology. However, these measurements have one important limitation: they contain little information about the spatial distribution of shape changes across the organism.

Landmark-based Geometric Morphometrics

In landmark-based geometric morphometrics that spatial information is contained in the data because the data are coordinates of "landmarks": discrete anatomical loci that are arguably homologous in all individuals in the analysis. For example, where sutures intersect is a landmark as are intersections between veins on an insect wing or leaf. So are foramina, small holes through which veins and blood vessels pass.

These points can be regarded as the "same" point in all specimens in the study. Landmark-based studies have traditionally analyzed 2D data, but with the increasing availability of 3D imaging techniques, 3D analysis are becoming more feasible even for small structures such as teeth. Finding enough landmarks to provide a comprehensive description of shape can be difficult when working with fossils or easily damaged specimens. That is because all landmarks must be present in all specimens, although coordinates of missing landmarks can be estimated. The data for each individual consists of a *configuration* of landmarks.

There are three recognized categories of landmarks. Type 1 landmarks are locally defined, meaning that they are defined in terms of structures close to that point; for example, an intersection between three sutures, or intersections between veins on an insect wing are locally defined and surrounded by tissue on all sides.

Type 3 landmarks, in contrast, are defined in terms of points far away from the landmark, and are often defined in terms of a point "furthest away" from another point. Type 2 landmarks are intermediate; this category includes points such as the tip structure, or local minima and maxima of curvature. They are defined in terms of local features, but they are not surrounded on all sides. In addition to landmarks, there are "semilandmarks," points along a curve. Their position along the curve is arbitrary but these points provide information about curvature in two or three dimensions.

Chapter 2

Spatial Analysis of Remote Sensing

In statistics, spatial analysis or spatial statistics includes any of the formal techniques which study entities using their topological, geometric, or geographic properties. The phrase properly refers to a variety of techniques, many still in their early development, using different analytic approaches and applied in fields as diverse as astronomy, with its studies of the placement of galaxies in the cosmos, to chip fabrication engineering, with its use of 'place and route' algorithms to build complex wiring structures. The phrase is often used in a more restricted sense to describe techniques applied to structures at the human scale, most notably in the analysis of geographic data. The phrase is even sometimes used to refer to a specific technique in a single area of research, for example, to describe geostatistics.

The history of spatial analysis starts with early cartography, surveying and geography at the beginning of history, although the techniques of spatial analysis were not formalized until the later part of the twentieth century. Modern spatial analysis focuses on computer based techniques because of the large amount of data, the power of modern statistical and geographic information science (GIS) software, and the complexity of the computational modelling. Spatial analytic techniques have been developed in geography, biology, epidemiology, sociology, demography, statistics, geoinformatics, computer science, mathematics, and scientific modelling.

Complex issues arise in spatial analysis, many of which are neither clearly defined nor completely resolved, but form the basis for current research. The most fundamental of these is the problem of defining the spatial location of the entities being studied. For example, a study on human health could describe the spatial position of humans with a point placed where they live, or with a point located where they work, or by using a line to describe their weekly trips; each choice has dramatic effects on the techniques which can be used for the analysis and on the conclusions which can be obtained. Other issues in spatial analysis include the limitations of mathematical knowledge, the assumptions required by existing statistical techniques, and problems in computer based calculations.

Classification of the techniques of spatial analysis is difficult because of the large number of different fields of research involved, the different fundamental approaches which can be chosen, and the many forms the data can take.

The History of Spatial Analysis

Spatial analysis can perhaps be considered to have arisen with the early attempts at cartography and surveying but many fields have contributed to its rise in modern form. Biology contributed through botanical studies of global plant distributions and local plant locations, ethological studies of animal movement, landscape ecological studies of vegetation blocks, ecological studies of spatial population dynamics, and the study of biogeography.

Epidemiology contributed with early work on disease mapping, notably John Snow's work mapping an outbreak of cholera, with research on mapping the spread of disease and with locational studies for health care delivery. Statistics has contributed greatly through work in spatial statistics. Economics has contributed notably through spatial econometrics. Geographic information system is currently a major contributor due to the importance of geographic software in the modern analytic toolbox. Remote sensing has contributed extensively in morphometric and clustering analysis. Computer science has contributed extensively through the study of algorithms, notably in computational geometry. Mathematics continues to provide the fundamental tools for analysis and to reveal the complexity of the spatial realm, for example, with recent work on fractals and scale invariance. Scientific modelling provides a useful framework for new approaches.

Fundamental Issues in Spatial Analysis

Spatial analysis confronts many fundamental issues in the definition of its objects of study, in the construction of the analytic operations to be used, in the use of computers for analysis, in the limitations and particularities of the analyses which are known, and in the presentation of analytic results. Many of these issues are active subjects of modern research.

Common errors often arise in spatial analysis, some due to the mathematics of space, some due to the particular ways data are presented spatially, some due to the tools which are available. Census data, because it protects individual privacy by aggregating data into local units, raises a number of statistical issues. Computer software can easily calculate the lengths of the lines which it defines but these may have no inherent meaning in the real world, as was shown for the coastline of Britain.

These problems represent one of the greatest dangers in spatial analysis because of the inherent power of maps as media of presentation. When results are presented as maps, the presentation combines the spatial data which is generally very accurate with analytic results which may be grossly inaccurate. Some of these issues are discussed at length in the book *How to Lie with Maps.*

Spatial Characterization

The definition of the spatial presence of an entity constrains the possible analysis which can be applied to that entity and influences the final conclusions that can be reached. While this property is fundamentally true of all analysis, it is particularly important in spatial analysis because the tools to define and study entities favor specific characterizations of the entities being studied.

Statistical techniques favor the spatial definition of objects as points because there are very few statistical techniques which operate directly on line, area, or volume elements. Computer tools favor the spatial definition of objects as homogeneous and separate elements because of the limited number of database elements and computational structures available, and the ease with which these primitive structures can be created.

There may also be arbitrary effects introduced by the spatial bounds or limits placed on a phenomenon within a study area. This occurs since spatial phenomena may be unbounded or have ambiguous transition zones. Spatial variables might be represented as a set of points or areas.

Oversimplification can occur in a GIS when data is represented as points but actually refer to areas, or visa versa. For example a pixel in a satellite image is an area represented by a point. Often environmental variables such as the earth surface temperatures, important in analyzing climate adaptation, are measured at points then interpolated into areas. The process of spatial sampling can sometimes involve ecological fallacy, atomic fallacy, or MAUP. A possible solution is a sensitivity analysis strategy using mutiple sets of spatial characterizations. It is also feasible to eliminate edge effects in spatial modelling and simulation by mapping the region to a boundless object such as a torus or sphere; or to use geographically weighted regression.

Spatial Dependency or Auto-correlation

A fundamental concept in geography is that nearby entities often share more similarities than entities which are far apart. This idea is often labelled 'Tobler's first law of geography' and may be summarized as "everything is related to everything else, but near things are more related than distant things". The opposite of spatial dependency is complete spatial randomness.

Spatial dependency is the co-variation of properties within geographic space: characteristics at proximal locations appear to be correlated, either positively or negatively. Spatial dependency leads to the spatial autocorrelation problem in statistics since, like temporal autocorrelation, this violates standard statistical techniques that assume independence among observations. For example, regression analyses that do not compensate for spatial dependency can have unstable parameter estimates and yield unreliable significance tests. Spatial regression models capture these relationships and do not suffer from these weaknesses. It is also appropriate to view spatial dependency as a source of information rather than something to be corrected.

Locational effects also manifest as spatial heterogeneity, or the apparent variation in a process with respect to location in geographic space. Unless a space is uniform and boundless, every location will have some degree of uniqueness relative to the other locations. This affects the spatial dependency relations and therefore the spatial process. Spatial heterogeneity means that overall parameters estimated for the entire system may not adequately describe the process at any given location.

This is a key concept for scientific analysis because most variables in the social and environmental sciences are spatially autocorrelated. For example, a contagious pathogen can sweep across spacetime as did the Black Death. A violent mentality can spread in a spatially autocorrelated fashion as it did in the Rwandan Genocide. Religious belief can propagate from place to place as did Buddhism, Christianity and Islam. Even scientific logic itself is believed to have rippled spacetime during the Scientific Revolution in an autocorrelated fashion. When the transmission vector is directly measurable as are pathogens, then there is evidence for a causal spatial correlation. When a transmission vector is less evident, unmeasurable or indirect (as in the case of feelings, beliefs and emotions) then the name changes to "spatial interaction".

Complete Spatial Randomness

Another fundamental concept in geography is the antithesis of spatial dependency, complete spatial randomness. CSR is the opposite of 'Tobler's first law of geography', situations where near things are not related. Stationarity is a fundamental assumption of traditional statistical hypothesis testing.

Scaling

Spatial measurement scale is a persistent issue in spatial analysis, more detail is available at the MAUP topic entry. Landscape ecologists developed a series of scale invariant metrics for aspects of ecology that are fractal in nature. In more general terms, no scale independent method of analysis is widely agreed upon for spatial statistics.

Sampling

Spatial sampling involves determining a limited number of locations in geographic space for faithfully measuring phenomena that are subject to dependency and heterogeneity. Dependency suggests that since one location can predict the value of another location, we do not need observations in both places. But heterogeneity suggests that this relation can change across space, and therefore we cannot trust an observed degree of dependency beyond a region that may be small. Basic spatial sampling schemes include random, clustered and systematic. These basic schemes can be applied at multiple levels in a designated spatial hierarchy (e.g., urban area, city, neighbourhood). It is also possible to exploit ancillary data, for example, using property values as a guide in a spatial sampling scheme to measure educational attainment and income. Spatial models such as autocorrelation statistics, regression and interpolation can also dictate sample design.

Common Errors in Spatial Analysis

The fundamental issues in spatial analysis lead to numerous problems in analysis including bias, distortion and outright errors in the conclusions reached. These issues are often interlinked but various attempts have been made to separate out particular issues from each other.

Length

In a paper by Benoit Mandelbrot on the coastline of Britain it was shown that it is inherently nonsensical to discuss certain spatial concepts despite an inherent presumption of the validity of the concept. Lengths in ecology depend directly on the scale at which they are measured and experienced. So while surveyors commonly measure the length of a river, this length only has meaning in the context of the relevance of the measuring technique to the question under study.

Locational Fallacy

The locational fallacy refers to error due to the particular spatial characterization chosen for the elements of study, in particular choice of placement for the spatial presence of the element. Spatial characterizations may be simplistic or even wrong. Studies of humans often reduce the spatial existence of humans to a single point, for instance their home address. This can easily lead to poor analysis, for example, when considering disease transmission which can happen at work or at school and therefore far from the home. The spatial characterization may implicitly limit the subject of study. For example, the spatial analysis of crime data has recently become popular but these studies can only describe the particular kinds of crime which can be described spatially. This leads to many maps of assault but not to any maps of embezzlement with political consequences in the conceptualization of crime and the design of policies to address the issue.

Atomic Fallacy

This describes errors due to treating elements as separate 'atoms' outside of their spatial context.

Ecological Fallacy

The ecological fallacy describes errors due to performing analyses on aggregate data when trying to reach conclusions on the individual units. Errors occur in part from spatial aggregation. For example a pixel represents the average surface temperatures within an area. Ecological fallacy would be to assume that all points within the area have the same temperature. This topic is closely related to the modifiable areal unit problem.

Modifiable Areal Unit Problem

The modifiable areal unit problem (MAUP) is an issue in the analysis of spatial data aggregated by zones. The conclusion of a statistical analysis depends on the particular shape or size of the zones used. It is a vexing and important issue because, the aggregation of point data into zones of different shapes and sizes can lead to opposite conclusions. The possibility of different, even contradictory answers, derived from one set of sampled data is an epistemic issue which affects our belief in the truth of objective analysis. More detail is available at the topic entry.

Solutions to the Fundamental Issues

Geographic Space: A mathematical space exists whenever we have a set of observations and quantitative measures of their attributes. For example, we can represent individuals' income or years of education within a coordinate system where the location of each individual can be specified with respect to both dimensions. The distances between individuals within this space is a quantitative measure of their differences with respect to income and education. However, in spatial analysis we are concerned with specific types of mathematical spaces, namely, geographic space. In geographic space, the observations correspond to locations in a spatial measurement framework that captures their proximity in the real world.

The locations in a spatial measurement framework often represent locations on the surface of the Earth, but this is not strictly necessary. A spatial measurement framework can also capture proximity with respect to, say, interstellar space or within a biological entity such as a liver. The fundamental tenet is Tobler's First Law of Geography: if the interrelation between entities increases with proximity in the real world, then representation in geographic space and assessment using spatial analysis techniques are appropriate. The Euclidean distance between locations often represents their proximity, although this is only one possibility. There are an infinite number of distances in addition to Euclidean that can support quantitative analysis. For example, "Manhattan" (or "Taxicab") distances where movement is restricted to paths parallel to the axes can be more meaningful than Euclidean distances in urban settings. In addition to distances, other geographic relationships such as connectivity (e.g., the existence or degree of shared borders) and direction can also influence the relationships among entities. It is also possible to compute minimal cost paths across a cost surface; for example, this can represent proximity among locations when travel must occur across rugged terrain.

Types of Spatial Analysis

Spatial data comes in many varieties and it is not easy to arrive at a system of classification that is simultaneously exclusive, exhaustive, imaginative, and satisfying. — G. Upton & B. Fingelton

Spatial Autocorrelation

Spatial autocorrelation statistics measure and analyse the degree of dependency among observations in a geographic space. Classic spatial autocorrelation statistics include Moran's *I* and Geary's *C*. These require measuring a spatial weights matrix that reflects the intensity of the geographic relationship between observations in a neighbourhood, e.g., the distances between neighbours, the lengths of shared border, or whether they fall into a specified directional class such as "west." Classic spatial autocorrelation statistics compare the spatial weights to the covariance relationship at pairs of locations. Spatial autocorrelation that is more positive than expected from random indicate the clustering of similar values across geographic space, while significant negative spatial autocorrelation indicates that neighboring values are more dissimilar than expected by chance, suggesting a spatial pattern similar to a chess board.

Spatial autocorrelation statistics such as Moran's *I* and Geary's *C* are global in the sense that they estimate the overall degree of spatial autocorrelation for a dataset. The possibility of spatial heterogeneity suggests that the estimated degree of autocorrelation may vary significantly across geographic space.

Local spatial autocorrelation statistics provide estimates disaggregated to the level of the spatial analysis units, allowing assessment of the dependency relationships across space. *G* statistics compare neighbourhoods to a global average and identify local regions of strong autocorrelation. Local versions of the *I* and *C* statistics are also available.

Spatial Interpolation

Spatial interpolation methods estimate the variables at unobserved locations in geographic space based on the values at observed locations. Basic methods include inverse distance weighting: this attenuates the variable with decreasing proximity from the observed location. Kriging is a more sophisticated method that interpolates across space according to a spatial lag relationship that has both systematic and random components. This can accommodate a wide range of spatial relationships for the hidden values between observed locations. Kriging provides optimal estimates given the hypothesized lag relationship, and error estimates can be mapped to determine if spatial patterns exist.

Spatial Regression

Spatial regression methods capture spatial dependency in regression analysis, avoiding statistical problems such as unstable parameters and unreliable significance tests, as well as providing information on spatial relationships among the variables involved.

Depending on the specific technique, spatial dependency can enter the regression model as relationships between the independent variables and the dependent, between the dependent variables and a spatial lag of itself, or in the error terms. Geographically weighted regression (GWR) is a local version of spatial regression that generates parameters disaggregated by the spatial units of analysis. This allows assessment of the spatial heterogeneity in the estimated relationships between the independent and dependent variables.

Spatial Interaction

Spatial interaction or "gravity models" estimate the flow of people, material or information between locations in geographic space. Factors can include origin propulsive variables such as the number of commuters in residential areas, destination attractiveness variables such as the amount of office space in employment areas, and proximity relationships between the locations measured in terms such as driving distance or travel time. In addition, the topological, or connective, relationships between areas must be identified, particularly considering the often conflicting relationship between distance and topology; for example, two spatially close neighbourhoods may not display any significant interaction if they are separated by a highway. After specifying the functional forms of these relationships, the analyst can estimate model parameters using observed flow data and standard estimation techniques such as ordinary least squares or maximum likelihood. Competing destinations versions of spatial interaction models include the proximity among the destinations (or origins) in addition to the origin-destination proximity; this captures the effects of destination (origin) clustering on flows. Computational methods such as artificial neural networks can also estimate spatial interaction relationships among locations and can handle noisy and qualitative data.

Simulation and Modelling

Spatial interaction models are aggregate and top-down: they specify an overall governing relationship for flow between locations. This characteristic is also shared by urban models such as those based on mathematical programming, flows among economic sectors, or bid-rent theory. An alternative modelling perspective is to represent the system at the highest possible level of disaggregation and study the bottom-up emergence of complex patterns and relationships from Behaviour and interactions at the individual level.

Complex adaptive systems theory as applied to spatial analysis suggests that simple interactions among proximal entities can lead to intricate, persistent and functional spatial entities at aggregate levels. Two fundamentally spatial simulation methods are cellular automata and agent-based modelling. Cellular automata modelling imposes a fixed spatial framework such as grid cells and specifies rules that dictate the state of a cell based on the states of its neighboring cells. As time progresses, spatial patterns emerge as cells change states based on their neighbours; this alters the conditions for future time periods.

For example, cells can represent locations in an urban area and their states can be different types of land use. Patterns that can emerge from the simple interactions of local land uses include office districts and urban sprawl. Agent-based modelling uses software entities (agents) that have purposeful Behaviour (goals) and can react, interact and modify their environment while seeking their objectives. Unlike the cells in cellular automata, agents can be mobile with respect to space. For example, one could model traffic flow and dynamics using agents representing individual vehicles that try to minimize travel time between specified origins and destinations. While pursuing minimal travel times, the agents must avoid collisions with other vehicles also seeking to minimize their travel times. Cellular automata and agent-based modelling are complementary modelling strategies. They can be integrated into a common geographic automata system where some agents are fixed while others are mobile.

Geographic Information Science and Spatial Analysis

Geographic information systems (GIS) and the underlying geographic information science that advances these technologies have a strong influence on spatial analysis. The increasing ability to capture and handle geographic data means that spatial analysis is occurring within increasingly data-rich environments. Geographic data capture systems include remotely sensed imagery, environmental monitoring systems such as intelligent transportation systems, and location-aware technologies such as mobile devices that can report location in near-real time. GIS provide platforms for managing these data, computing spatial relationships such as distance, connectivity and directional relationships between spatial units, and visualizing both the raw data and spatial analytic results within a cartographic context.

Geographic knowledge discovery (GKD) is the human-centred process of applying efficient computational tools for exploring massive spatial databases. GKD includes geographic data mining, but also encompasses related activities such as data selection, data cleaning and pre-processing, and interpretation of results. GVis can also serve a central role in the GKD process. GKD is based on the premise that massive databases contain interesting (valid, novel, useful and understandable) patterns that standard analytical techniques cannot find. GKD can serve as a hypothesis-generating process for spatial analysis, producing tentative patterns and relationships that should be confirmed using spatial analytical techniques. Spatial Decision Support Systems (sDSS) take existing spatial data and use a variety of mathematical models to make projections into the future. This allows urban and regional planners to test intervention decisions prior to implementation.

Geovisualization

Geovisualization, short for *Geographic Visualization*, refers to a set of tools and techniques supporting geospatial data analysis through the use of interactive visualization. Like the related fields of scientific visualization and information visualization geovisualization emphasizes knowledge construction over knowledge storage or information transmission. To do this, geovisualization communicates geospatial information in ways that, when combined with human understanding, allow for data exploration and decision-making processes.

Traditional, static maps have a limited exploratory capability; the graphical representations are inextricably linked to the geographical information beneath. GIS and geovisualization allow for more interactive maps; including the ability to explore different layers of the map, to zoom in or out, and to change the visual appearance of the map, usually on a computer display. Geovisualization represents a set of cartographic technologies and practices that take advantage of the ability of modern microprocessors to render changes to a map in real time, allowing users to adjust the mapped data on the fly.

History

The term visualization is first mentioned in the cartographic literature at least as early as 1953, in an article by University of Chicago geographer Allen K. Philbrick. New developments in the field of computer science prompted the National Science Foundation to redefine the term in a 1987 report which placed visualization at the convergence of computer graphics, image processing, computer vision, computer-aided design, signal processing, and user interface studies and emphasized both the knowledge creation and hypothesis generation aspects of scientific visualization. Geovisualization developed as a field of research in the early 1980s, based largely on the work of French graphic theorist Jacques Bertin. Bertin's work on cartographic design and information visualization share with the National Science

Foundation report a focus on the potential for the use of "dynamic visual displays as prompts for scientific insight and on the methods through which dynamic visual displays might leverage perceptual cognitive processes to facilitate scientific thinking". Geovisualization has continued to grow as a subject of practice and research. The International Cartographic Association (ICA) established a Commission on Visualization & Virtual Environments in 1995.

Related Fields

Geovisualization is closely related to other visualization fields, such as scientific visualization and information visualization. Owing to its roots in cartography, geovisualization contributes to these other fields by way of the map metaphor, which "has been widely used to visualize non-geographic information in the domains of information visualization and domain knowledge visualization. It is also related to urban simulation.

Practical Applications

Geovisualization has made inroads in a diverse set of real-world situations calling for the decision-making and knowledge creation processes it can provide. The following list provides a summary of some of these applications as they are discussed in the geovisualization literature.

Forestry

Geovisualizers, working with European foresters, used CommonGIS and Visualization Toolkit (VTK) to visualize a large set of spatio-temporal data related to European forests, allowing the data to be explored by non-experts over the Internet. The report summarizing this effort "uncovers a range of fundamental issues relevant to the broad field of geovisualization and information visualization research".

The research team cited the two major problems as the inability of the geovisualizers to convince the foresters of the efficacy of geovisualization in their work and the foresters' misgivings over the dataset's accessibility to non-experts engaging in "uncontrolled exploration". While the geovisualizers focused on the ability of geovisualization to aid in knowledge construction, the foresters preferred the information-communication role of more traditional forms of cartographic representation.

Archaeology

Geovisualization provides archaeologists with a potential technique for mapping unearthed archaeological environments as well as for accessing and exploring archaeological data in three dimensions.

The implications of geovisualization for archaeology are not limited to advances in archaeological theory and exploration but also include the development of new, collaborative relationships between archaeologists and computer scientists.

Environmental Studies

Geovisualization tools provide multiple stakeholders with the ability to make balanced environmental decisions by taking into account the "the complex interacting factors that should be taken into account when studying environmental changes". Geovisualization users can use a georeferenced model to explore a complex set of environmental data, interrogating a number of scenarios or policy options to determine a best fit.

Urban Planning

Both planners and the general public can use geovisualization to explore real-world environments and model 'what if' scenarios based on spatio-temporal data. While geovisualization in the preceding fields may be divided into two separate domains—the private domain, in which professionals use geovisualization to explore data and generate hypotheses, and the public domain, in which these professionals present their "visual thinking" to the general public—planning relies more heavily than many other fields on collaboration between the general public and professionals. Planners use geovisualization as a tool for modelling the environmental interests and policy concerns of the general public. Jiang et al. mention two examples, in which "3D photorealistic representations are used to show urban redevelopment [and] dynamic computer simulations are used to show possible pollution diffusion over the next few years." The widespread use of the Internet by the general public has implications for these collaborative planning efforts, leading to increased participation by the public while decreasing the amount of time it takes to debate more controversial planning decisions.

Digital Mapping

Digital mapping (also called digital cartography) is the process by which a collection of data is compiled and formatted into a virtual image. The primary function of this technology is to produce maps that give accurate representations of a particular area, detailing major road arteries and other points of interest. The technology also allows the calculation of distances from once place to another. Though digital mapping can be found in a variety of computer applications, such as Google Earth, the main use of these maps is with the Global Positioning System, or GPS satellite network, used in standard automotive navigation systems.

History

From Paper to Paperless: The roots of digital mapping lie within traditional paper maps such as the Thomas Guide. Paper maps provide basic landscapes similar to digitized road maps, yet are often cumbersome, cover only a designated area, and lack many specific details such as road blocks. In addition, there is no way to "update" a paper map except to obtain a new version. On the other hand, digital maps, in many cases, can be updated through synchronization with updates from company servers.

Expanded Capabilities

Early digital maps had the same basic functionality as paper maps—that is, they provided a "virtual view" of roads generally outlined by the terrain encompassing the surrounding area. However, as digital maps have grown with the expansion of GPS technology in the past decade, live traffic updates, points of interest and service locations have been added to enhance digital maps to be more "user conscious." Traditional "virtual views" are now only part of digital mapping. In many cases, users can choose between virtual maps, satellite (aerial views), and hybrid (a combination of virtual map and aerial views) views. With the ability to update and expand digital mapping devices, newly constructed roads and places can be added to appear on maps.

Data Collection

Digital maps heavily rely upon a vast amount of data collected over time. Most of the information that comprise digital maps is the culmination of satellite imagery as well as street level information. Maps must be updated frequently to provide users with the most accurate reflection of a location. While there is a wide spectrum on companies that specialize in digital mapping, the basic premise is that digital maps will accurately portray roads as they actually appear to give "life-like experiences."

Functionality and Use

Computer Applications: Computer programs and applications such as Google Earth and Google Maps provide map views from space and street level of much of the world. Used primarily for recreational use, Google Earth provides digital mapping in personal applications, such as tracking distances or finding locations.

Scientific Applications

The development of mobile computing has recently (since about 2000) spurred the use of digital mapping in the sciences and applied sciences. As of 2009, science fields that use digital mapping technology include geology, engineering, architecture, land surveying, mining, forestry, environmental, and archaeology.

GPS Navigation Systems

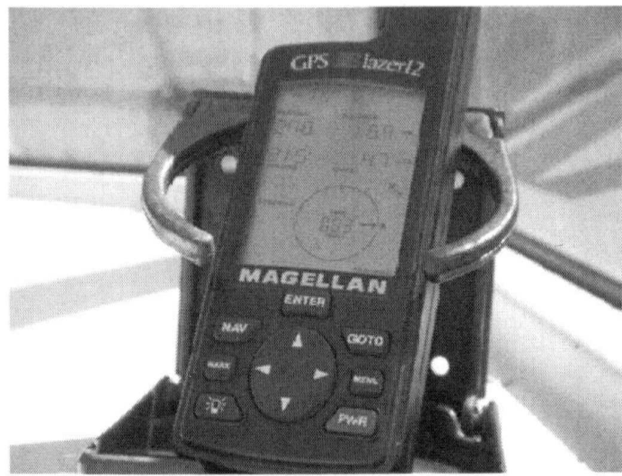

Figure: *Civilian GPS receiver ("GPS navigation device") in a marine application.*

The principle use by which digital mapping has grown in the past decade has been its connection to Global Positioning System (GPS) technology. GPS is the foundation behind digital mapping navigation systems.

How It Works

Figure: *Automotive navigation system in a taxicab.*

The coordinates and position as well as atomic time obtained by a terrestrial GPS receiver from GPS satellites orbiting Earth interact together to provide the digital mapping programming with points of origin in addition to the destination points needed to calculate distance.

This information is then analysed and compiled to create a map that provides the easiest and most efficient way to reach a destination.

More technically speaking, the device operates in the following manner:

1. GPS receivers collect data from "at least twenty-four GPS satellites" orbiting the Earth, calculating position in three dimensions.
2. The GPS receiver then utilizes position to provide GPS coordinates, or exact points of latitudinal and longitudinal direction from GPS satellites.
3. The points, or coordinates, output an accurate range between approximately "10-20 meters" of the actual location.
4. The beginning point, entered via GPS coordinates, and the ending point, (address or coordinates) input by the user, are then entered into the digital map.
5. The map outputs a real-time visual representation of the route. The map then moves along the path of the driver.
6. If the driver drifts from the designated route, the navigation system will use the current coordinates to recalculate a route to the destination.

Geospatial

Geospatial is a term widely used to describe the combination of spatial software and analytical methods with terrestrial or geographic datasets. The term is often used in conjunction with geographic information systems and geomatics, never separately. Geospatial Information Officer is the head of geospatial information technology within a civilian, business, government and military organizations.

The term "geospatial" is an adjective that means, "of or relating to the relative position of things on the earth's surface." The term geospatial can further be defined as, " pertaining to the geographic location and characteristics of natural or constructed features and boundaries on, above, or below the earth's surface; esp. referring to data that is geographic and spatial in nature."

Many Geographic information system (GIS) products apply the term geospatial analysis in a very narrow context. In the case of vector-based GIS this typically means operations such as map overlay (combining two or more maps or map layers according to predefined rules), simple buffering (identifying regions of a map within a specified distance of one or more features, such as towns, roads or rivers) and similar basic operations.

This reflects (and is reflected in) the use of the term spatial analysis within the Open Geospatial Consortium (OGC) "simple feature specifications". For raster-based GIS, widely used in the environmental sciences and remote sensing, this typically means a range of actions applied to the grid cells of one or more maps (or images) often involving filtering and/or algebraic operations. These techniques involve processing one or more raster layers according to simple rules resulting in a new map layer, for example replacing each cell value with some combination of its neighbours' values, or computing the sum or difference of specific attribute values for each grid cell in two matching raster datasets. Descriptive statistics, such as cell counts, means, variances, maxima, minima, cumulative values, frequencies and a number of other measures and distance computations are also often included in this generic term spatial analysis.

However, the above list of functions cover only the most basic of facilities, albeit those that may be the most frequently used by the greatest number of GIS professionals. To this initial set must be added a large variety of statistical techniques (descriptive, exploratory, and explanatory statistics) that have been designed specifically for spatial and spatio-temporal data. Today such techniques are of great importance in the social sciences, medicine and criminology, despite the fact that their origins may often be traced back to problems in the environmental and life sciences, in particular ecology, geology and epidemiology. It is also to be noted that spatial statistics is largely an observational science (like astronomy) rather than an experimental science (like agronomy or pharmaceutical research). This aspect of geospatial science has important implications for analysis, particularly the application of a range of statistical methods to spatial problems.

Limiting the definition of geospatial analysis to 2D mapping operations and spatial statistics remains too restrictive. There are other very important areas to be considered. These include: surface analysis —in particular analysing the properties of physical surfaces, such as gradient, aspect and visibility, and analysing surface-like data "fields"; network analysis — examining the properties of natural and man-made networks in order to understand the behaviour of flows within and around such networks; and locational analysis. GIS-based network analysis may be used to address a wide range of practical problems such as route selection and facility location (core topics in the field of operations research, and problems involving flows such as those found in hydrology and transportation research. In many instances location problems relate to networks and as such are addressed with tools designed for this purpose, but in others existing networks may have little or no relevance or may be impractical to incorporate within the modelling process. Problems that are not specifically network constrained, such as new road or pipeline routing, regional warehouse location, mobile phone mast positioning or the selection of rural community health care sites, may be effectively analysed (at least initially) without reference to existing physical networks. Locational analysis "in the plane" is also applicable where suitable network datasets are not available, or are too large or expensive to be utilised, or where the location algorithm is very complex or involves the examination or simulation of a very large number of alternative configurations.

A further important aspect of geospatial analysis is geovisualization — the creation and manipulation of images, maps, diagrams, charts, 3D views and their associated tabular datasets. GIS packages increasingly provide a range of such tools, providing static or rotating views, draping images over 2.5D surface representations, providing animations and fly-throughs, dynamic linking and brushing and spatio-temporal visualisations. This latter class of tools is the least developed, reflecting in part the limited range of suitable compatible datasets and the limited set of analytical methods available, although this picture is changing rapidly. All these facilities augment the core tools utilised in spatial analysis throughout the analytical process (exploration of data, identification of patterns and relationships, construction of models, and communication of results).

Computerised Cadastral Mapping

Applications of cartography in rural mapping are age old and this art has taken various turns since from its existence till today. The new trends in cartographic development has endeavor this art with digital techniques. The making and use of computer based systems for the cartography is added advantage in reproduction and multidisciplinary use and analysis of thematic mapping in the rural development activities. In fact this transition to "computer cartography" started long time ago. It is worth noting, however, that the pace of change towards an electronic future is variable across the world. Digital cadastral maps are being produces as cartographic maps, which provide an overview of some of the major trends and concerns in digital cartography. The paper is attributed the contribution of applications of CAD (Computer Aided Drafting) in digital cartography in addition to Geographical Information Systems applications for thematic mapping through digital analysis. Digital cadastral maps are used as a media for generating and preparation query cell for rural planning as well.

Every activity needs proper planning, reporting and monitoring. A well-formed information base is most essential for support during these processes. The graphical form of data is most easily understood at any level of functionary and is most properly interpreted, and useful actions can be followed.

As far as storage and processing of information is concerned, computers are already quite accepted tool. Computers mainly help in storing information in the form of maps and help the user in performing complex tasks in such simplified way. Sketchy maps and illustrations used to record land ownership and revenue information in India, later gave rise to cadastral maps. Assessment of crop yield and fixing the governmental share became a major activity of revenue administration. Scale alteration and superimposition are hard exercises if one is working with printed maps (on paper), but the same are easily performed through digital maps.

Beginning in the 1970's, many mapping experts adopted a communications model for cartography, understanding maps as tools for the communication of information from cartographer to map user. With the rapid progress in computer technology afforded by the ubiquitous personal computer, in the last decade a number of cartographic researchers, led by Alan MacEachren (1995), have suggested a new way of understanding how maps work. Rather than attempting to make a best map, modern computer technology can allow for the preparation of multiple representations of a phenomenon that can be used to answer different questions.

High-resolution digital data (satellite image/aerial photographs) can be used for capturing the individual land holdings. It also aids in easy retrieval and manipulation of data, to update new land parcels to enable periodic report generation and its changing pattern.

Objectives

Objectives of this paper is to highlight the needs of the digital cadastral mapping with cadastral information at larger scales for physical planning and development of rural areas and to drive the such need to the survey and mapping community.

Use of Large-scale Cadastral Maps

Preparation and use of large-scale maps, especially for rural areas, is not as good as in developed and other developing countries (Rudraiah). Maps are required by many department, NGOs, agencies, companies to carry out developmental activities in the area. Local authorities, public undertakings, service organisation require maps. However, the requirement of maps in terms of quantity, quality and accuracy vary from organisation to organisation. It is important to note that all the agencies aforementioned and others do not need comprehensive map, i.e. all the information in map.

Need For Computerised Village Maps

Cartographers use methods of showing objects on maps in various ways- for example symbols or objects, lines, hatching boundaries etc. Usually separate maps are created for different subjects or themes, thus, giving rise to numerous maps of the same area. Since in most cases these maps are produced by different agencies using no common methodology, the scales of these maps are also quite different. Computerised maps comes very handy for these initial stages of mapping and prove much more useful when mapped data is put to use. Computerised maps allow mapping of details and minor alterations wherever required. The selected superimposition of selected themes gives idea of an integrated picture and helps to understand area by considering variety of features together.

Lacunae in Existing Cadastral Maps

Exhaustive information is available but hardly used. The information collected from the revenue departments is exhaustive but not used at par. This ignores the logical information principle that as far as possible collection of information should be avoided if it is not going to be used. Further, a number of registers need to be unnecessarily maintaining the auxiliary information. Hard print cadastral maps are not easily available for planning process and area away from the reach of common people. Such maps are only the property of government authority and are used as desired by them.

Development in Digital Cartography

The move towards automation has necessitated some redesigning of maps by eliminating some content (Hadley, 1987) and modifications of others like the conventional hatching replaced by coloured filling of specific symbol.

Simplification of information and graphic design are often undertaken simultaneously in manual generalisation, in digital cartography there is a need to distinguish between information generalisation and display generalisation; the former contributes to digital mapping, the latter to visual mapping. Both types of transformations are intellectually demanding and involve a variety of subtasks and processes. Many applications use maps of the same entities compiled at a variety of scales with different content and symbolism. Many land information systems focus on the land parcel and are concerned with the areal extent of such linear objects. A number of users, such as in local government and NGOs, need both detailed and simplified representations of the same objects. At present, scale-related requirements are met by multiple definitions of the same areas. The scale of display appropriate for each feature is also recorded within a scale-integrated database. Systems based on raster/video images zoom by retrieving images captured at the next larger scale. Other researchers are manually integrating road networks captured at different scales for route planning and are investigating database designs for supporting scale-free mapping (Abraham, 1989). The digital mapping has completely transformed cartography since the mid 1980s where map output in fast multiple reproduction made easier.

Methodology

Monitoring is exercised to make a definite assessment of results of actions and activities and there by decide if any change required. Monitoring through digital maps is very effective and useful. By pointing initial and current status on maps the impact area may be easily seen. Creating maps, union, intersection, subtraction and analyses are easily carried out.

Computerization: There is a considerable effort in culling out the information from the registers to prepare reports in the current manual system. The load of processing/calculation is currently placed on lower staff to maintain documents. The objective is to map out land parcel wise basic indicators/vital statistics. It is envisaged that all data would be computerised and presented in a visual form on a digitised map.

After the initial exercise, including necessary amendments to methodology; it is decided to undertake the task for cadastral mapping. The mapping involved the following two processes being undertaken simultaneously:

1. Computerised digitization of maps
2. Data collection and updation.

Both processes converged once they were individually completed-i.e. the land parcel-wise data was entered into tabular fields that were linked to each land parcel on the digital map.

Computerised digitization of maps and data is undertaken using cadastral maps basic data like land parcel size, ownership, landuse, cropping pattern and infrastructure details obtained from the "Patwari" as a base. Basic physical features, boundaries and important landmarks were digitized. Other primary baseline and temporal data collected entered in computer system the formulate database.

Cadastral Mapping

Village Khasra (Cadastral) map is very vital and important because it carries the smallest ever mapped land unit in it. The villagers use the maps for general observation at village level and useful for macro level application. Cadastral maps are generally available on 1:6000 or 1:4000 scales. The cadastral maps are consists of all three types of geometries, i.e. point, line and area (polygon). In GIS based cadastral maps all the characteristics like geometry, attribute and map objects are present.

The cadastral maps are generalized form of real earth features on a paper, which involves the application of processes, such as selection, classification, simplification, symbolization and pattern recognition. All cadastral maps are symbolic and rely on the use of a graphic sign language.

All the feature, linear and aerial, are demarcated at real scale and area and length of the linear features such as roads and rivers marked by solid double lines can be measures with accuracy due to large scale mapping. While in small scale mapping, such as topographical maps show lengths of linear feature on its scale but areas of most linear features such as roads, railways and single line drainage are as representatives only (Singh, 2005). The design of the symbolization system and of its structure is been based on different parameters. For example, one well within the cadastral map is shown as a small circle, and further qualitative differentiation, based on the lined and unlined well, is also made possible using different patterns of circle.

Cadastral Resource Mapping

Present important resources available at land parcel level can be considered real perspective for planning purpose. A village resource map gives a useful scenario of the present available facilities & resources, which helps in identification of further required facilities. Therefore, cadastral map used for resource mapping.

Cadastral Database

A database is useless unless populated with appropriate data and kept up to date with temporal changes. Digital mapping has a strong capacity to integrate and cover the full range of data sources. The database structure with special reference to Madhya Pradesh cadastral maps can have following records, which are to be constantly updated (Singh, 2005). The record of each map includes information about its:

Title Information		*Records*	
1.	Name of Village	1.	Khasra Number (Land Parcel No.)
2.	Name of Tehsil	2.	Area in Acres/Hectares (description)
3.	Name of District	3.	Ownership/Partners/Father's/ Husband's name with place of residence/revenue tax to be paid.
4.	Mouja Number	4.	Name of Patta Holder etc.
5.	Halka Number	5.	Name of Crop
6.	Sheet number	6.	Area sown
7.	Year of preparation	7.	Double crop land
8.	Year of updation	8.	Current Fallow Land
		9.	Fallow Land for 2 to 5 Years
		10.	Fallow Land for more than 5 years
		11.	Name of crop and area sow out of owned land.
		12.	Remarks (details regarding other amenities like well, tube well, trees etc.)

Components of Digital of Cadastral Maps

The digital cartography involves many components to be decided at first hand. Following the major components are used in it:

Map Layers

The layers can be assigned them to different objects in the map to identify different objects easily. Different layers can be assigned different colours and line types. The layers can control the visibility, colour and line type of the objects on a layer for editing and analysis. One object can be moved from one layer to another and even one can change the name of a layer. Layers are the different overlays used in paper-based drafting. The layers are the primary organizational tools in the CAD.

Symbolization

Symbolization is the science of signs assigned to particular function, with sign considered to be a relationship between an expression and its referent. For cartography, maps can be viewed as tools to communicate meaning primarily through symbols. This implies that there must be a cartographic language based on symbols that provides a map with meaning. Symbolization is most important part of cartography. In the present paper a rich combination of colour, pattern, shape and size of symbols is provided according to their functions for defining and applying symbolization styles to point, line, or area data. In digital cartography vector and bit mapped symbols are used.

List of Symbols

Apart from the identification of symbolic feature following parameters plays an important role in symbolization process:

1. Location
2. Size
3. Colour (Value)
4. Texture
5. Colour (Hue)
6. Orientation
7. Shape.

Text and Annotation

Digital cartography has significant capabilities for text labeling. Many text font types are now supporting it. Font information can now be embedded in output formats. There are numerous capabilities for storing text in the database as annotations. Annotations are well liked with different font formats so that they can be automatically updated as change happens in the real world. Annotation support now includes such cartographic elements as callouts and leader lines. The fonts provide the cartographer with the freedom to override the strict rules of the font selection. The stored representation capabilities of geo database facilitate to derive, create, and store cartographic features as layers of points, lines, or polygons. Such cartographic feature classes are often vital to the implementation of digital cartographic models from digital landscape models and can act as a source for subsequent stored representations.

Based on the requirement of different departments and NGOs and approval of authorities extensibility of digital cadastral mapping can be provided to make the Land Records easy to use. Thus, creating a transparent administrative environment. Most State Land Records Departments could use the methodology to make it usual practice.

The rapid geographic information technology (GIT) development and the adapting of GIT tools and methodology can be noted in many different areas of human beings interests from physical geography to social networks theory. These technological trends have been bringing certain changes in geographic, or spatial, data retrieval, visualization, and, what is the most important, – data analysis and data interpretation.

Significant role in the process of geographical data use transformation played geographical information systems (GIS) – software, hardware, human resources, georeferenced data, and models. Because linking location to information is a process that applies to many aspects of decision making, GIS has contributed to many scientific disciplines growth, including but not limited to cartography, computer science, and engineering.

However, even due to appearance different Internet-based services like Google Maps and Google Earth that have integrated a huge amount of spatial data which allow to use these resources for wide audience (Jiang, 2007) and free access to amount of spatial data in a form of satellite imageries data, e.g., Landsat, Quickbird satellites, geographical data stays one of the most complicated data from the data structure point of view (i.e. spatial data is massive, temporal and spatial referenced; geographic features consist of geometry, optionally attributive part in a relational form in several dimensions; GIS data includes information about spatial relationships (topology), etc.) In addition, spatial data analysis is a very complicated process due to its complexity: data exploration process assumes completeness of the data, but in the reality the data is fuzzy and inaccurate because as a rule it is collected from different source and errors and incompleteness are inherent. This way, GIS differs from any other informational systems that deal with relational or even object-oriented data structure models. Another critical moment in the process of cartography as the technology of geographic data visualization and GIS as a framework of spatial data collection and processing integration is related to the schema of interaction with data. Thus, all schema parts of the data mining process have been replaced. And if the processes of raw data collection (by using GPS, satellite cameras or by field surveying means), data organization (scanning, vectorization, normalization, abstraction) are renovated by cutting edge technologies in some way, so the final step in the process of data mining, that is, data exploration (spatial analysis, modelling), has been changed totally and the rate of appearance of new ways of data visualization and map-user interaction is amazing.

Along with the increasing range and sophistication of spatial analysis facilities that can be found within commercial GIS software, there has been a major transformation in spatial analytical techniques. In large, this has come about as a result of technological development and the related availability of software tools and access to high-quality publicly available sets of data. One aspect of this that can be depicted already — the shift towards network-based location modelling where in the past this would have been unfeasible. More general shifts can be also seen in the move towards local rather than simply global analysis, for example in the field of data research, and in the development of a wide range of modelling methods that handle problems through micro-scale processes by using geocomputational methods. This way, emerging technologies like Internet and Web influence the way of carrying out researches and certainly will play a major role in the next generation of GIS aspects of data modelling, processing, formalization, analysis, and data distribution.

In a more ample treatment, this integration of GIS techniques of geographic data exploration and spatial analysis has brought tremendous changes in the process of transformation geographical data to information and then to the geographic knowledge and links GIS, maps and other advanced computer visual data representations methods together. In addition, while integrating on the scientific level, geographic information technology and science is permanently developing, extending, and changing on the productive level as well. One of the additional important substitution is related to the process of data discovery: today the impact of geographic phenomena or event on humanity in a post- spatial temporal context investigation becomes more vital task and it can be concluded that a lot of efforts will be put to invent new ways of effective data visualization that will allow to discover geographical trends between real world objects in many dimensions.

It should be noted that such analytical and visual tools integration will bring absolutely new ways of wide users and map interaction. That means that data presentations are moving from 2D and 3D visualization to 4D presentation. Nevertheless, earlier, spatial data correlation was defined within the exhibition of spatial autocorrelation when values measured nearby in space are more similar than values measured farther away from each other (Tobler, 2000). But now spatial data autocorrelation model can be extended by temporal data: it is possible to exhibit serial autocorrelation when values measured close together in time are more similar than values measured far apart in time. So, in the future temporal data structures will accommodate time data as a stored dimension and completely modified the conventional mapping paradigm.

All of these changes stipulate emerging of new methodology, scientific methods, and force standard fundamental approaches to adapt new ways of information usage. It can be proposed that such shifts will continue to alter the way of people thinking and communication; will add geographical, or spatial, context to many areas of human beings activities and hence, the fundamental approaches of geospatial data usage and the ways of using GIS in the future will be permanently changing.

Geomatics and Geoinformatics in Modern Information Society – projection of New Trends into their Curricula at the University of West Bohemia in Pilsen

Spectacular progress in electronics and computer science brought revolutionary changes of scientific and technical disciplines, as e.g. geodesy, cartography, photogrammetry, cadastral and topographic mapping, in the last 25 years. They influenced substantially the style of surveyor´s activities. Traditional tools for his work – survey tape, stakes, theodolite and paper maps have been replaced by new and powerful tools – electronic tacheometry, Global Positioning System, LIDAR, digital photogrammetry, digital cartography and geographic information systems. To-day´s primary products of surveying and mapping are digital databases of geospatial information which can be analysed, modelled and integrated with other kind of information and presented to the user in digital or graphical form tailored to his specific demands. To-day´s surveyor or cartographer cannot live in close shells of individual disciplines (applied or theoretical geodesy, topography, photogrammetry, cartography). His profession must become an integrated profession called in many countries geomatics or geospatial information engineering.

Even if this term is mixed up with the term geoinformatics in most Central- and East European countries there is a certain difference in contents what will be evident when comparing curriculums of both disciplines at some Czech universities or some official definitions. According to ISO Standard 19122 „geomatics is a discipline concerned with the collection, distribution, storage, analysis, processing, presentation of geographic data or geographic information". Its range is perfectly described by activities of the Geomatics Canada: establishing and maintainace of national spatial reference system, preparing, publishing and distributing of state topographical maps, aeronautical charts, aerial photographs and gazetteers, surveys on state boundaries, property surveys on federal lands, maintainance of national bases of geographic data for the development of geographical information systems.

There is no definition of geoinformatics in ISO Standards. One of the best was published by Dietmar Grunreich, president of the Federal Agency for Cartography and Geodesy in Frankfurt (Main): "geoinformatics is a discipline concerned with theory of geospatial data modelling, their storage, management and processing as well as with development of geographical information systems and necessary information and communication technology". Modern conception of geomatics was accepted by the University of West Bohemia in Pilsen in 1995 and projected into curriculum for geomatics. Other universities in the Czech Republic follow more or less the objectives of geoinformatics described by D. Grunreich.

Because of a great demand for graduates of geoinformatics the BSc and MSc curricula for geoinformatics were accreditated at the University of West Bohemia in Pilsen in 2006, too. Composition of these curricula is based on profound analysis of future trends and applications of geosciences which are tied in with geomatics and geoinformatics. It was also necessary to specify future legal and political environment in Europe that can influence substantionally the level of global, European and national infrastructure of geospatial data. In the next 10 years a general harmonization of political and legal environment in the European Union is expected. It enables to create the European geoinformation infrastructure which will be characterized by horizontal interoperability of national geographic databases and by hierarchical interoperability from the level of Land information systems up to national, continental and global geographic databases. Geodata will be generally accessible (e.g. through EU-Geoportal, cross-border geodata exchange will be obvious.

Development of e-cadastre of real estates will lead up to the sole information system which will register both ownership rights to real estates inclusive their restrictions, and technical data on owners, lands, buildings and flats. 3-D cadastre will also register and represent underground- and overhead objects of real estate character and locate them in national (later in European) coordinate reference system. Maintenance of national cadastral information systems will be centralized, but on contrary, data distribution to users will be spread into information kiosks of local administration. European Land Information Service (EULIS) will provide the electronic interface which enables the access to national cadastral information systems in Europe. Costs of their services will be fully covered with revenues coming from the fees.

Private surveyors (geomaticians) will be forced to provide the clients with complex services inclusive real estate evaluation and sale. Many stereotype surveying operations will be enough automated so that they will be carried out by people without professional education. As an example the application of robotized total stations for detailed surveys or the system of terrestrial laser scanning for surveying of buildings, streets, industrial objects, engineering constructions and underground objects may be introduced. Surveying and mapping will lead up to the integration of more scientific and technical disciplines (as geomatics or geospatial information engineering).

Development of theoretical geodesy will concentrate on improving the parameters of the Earth body and its gravity field as well as on precising the continental and national coordinate reference systems inclusive their temporal changes. Global positioning system will reach millimeter accuracy in geodetic kinematic applications, too. Dense network of permanent stations in individual countries, currently used by intellectual systems for navigation of persons, cars, ships and airplanes in the real time, will be their component. Temporal-space aspect will be typical for gathering, processing and analysis of geospatial data.

In photogrammetry the transformation of photography into digital image recording will be finalized. The price of large-format digital aerial cameras will be comparable with contemporary film cameras if not cheaper. Spatial position of individual digital images will be derived from directly measured elements of exterior orientation furnished by GPS and inertial measuring unit during the survey flight. Airborne multispectral digital images will be frequently used to various thematic applications. An airborne laser scanner (LIDAR) will be routinelly applied to modelling of terrain or surface relief and their temporal changes.

Remote sensing technologies will utilize many small satellites providing multispectral digital images with large ammount of narrow visible, infrared and thermal wavebands. High resolution satellite digital images will replace to some extent aerial photogrammetry in the case of updating the topographic databases. The methods of automated recognition of spatial and qualitative changes during the time period of updating will be frequently used for this reason.

Thanks to capacities and large field-of-view of remote sensing the whole Earth surface inclusive non-inhabited regions will be represented in the form of digital topographic databases at level of 1: 1 million, 1:250 thousand and even 1: 50 thousand mapping by the year 2025.

Cartography will be significantly influenced by the development of information technologies. It seems that it will be above all a service for cartographic visualization of geospatial data processed by GIS technologies. The map in paper form although will not loose its significance but it will be one of tools for education, free time activities and common military purposes. Other more frequent forms will be electronic maps and atlases, 3D models of landscape, animations, virtual models and intellectual geoimages of multimedia character. Internet will facilitate quick access to cartographic products and it becomes a global geoinformation system. Abandonment of cartographic know-how when creating the GIS software by IT- people only (e.g. principles of cartographic generalization, application of sofisticated map language) could lead up to distribution and using the impressive and quick working software but not always giving meaningful outputs.

Analysis of future trends and applications of geosciences which are tied in with geomatics and geoinformatics illustrates how deep will be changed the profession of to-day´s surveyor and cartographer in coming information age. Focus of their activities will transfer from labour-intensive collection of geospatial data to their processing, maintainance and presentation for the needs of scientific, administrative, legal and technical operations. The curricula for geomatics and geoinformatics should reflect above mentioned trends.

Remote Sensing and GIS- Application in Manipur

Applications in the Development Activities of Manipur

In a state like Manipur, where there are more hilly areas than plains and where there is the disadvantage of travel from one place to another to take up some development programme or to make an environmental survey, Remote Sensing has helped tremendously. Therefore, by taking the advantages of Remote Sensing, many developmental progresses can be made by the various government departments of Manipur. In a less developed state like Manipur, remote Sensing centre will give much benefit. A layman may not know what remote sensing is? In simple terms, it is identifying the objects at a place where we can not reach, without going at that place. For example, we can easily know whether there are trees growing or human beings settled at a place beyond many ranges of hills with the help of remote sensing.

Till now, with help of remote sensing, we have been able to know about the various natural resources on the earth's surface. In this modern age, with the advantages of remote sensing, there will be much more progress. To set an example, in the forest department, we can easily identify the place where trees are growing, where there is no growth and where we can grow trees and crops. In the field of agriculture too, we can know the crops grown at a particular place or which crops would be more suitable for growing at the place. To dig a canal, to make new roads at a suitable place, to identify water resources or to select a place for digging tube-well, all there can be easily done through remote sensing. Regarding how remote sensing works:- it is with the help of powerful cameras or sensors take photos of the surface of the earth and the required information can be taken from those photos. These cameras or sensors take the help of airplanes, balloons, helicopters or satellites as their platform. Nowadays, we have got sensors which can take a clear picture of very small objects, say about an area of 1 square meter.

Information technology, which we very often talk about, is also mainly the work of remote sensing. In our state also, in many government department, there is huge number of computer, it these computers are lacking information's then, it won't be wrong to say that those computers are simply typewriters. If a planner or a decision maker is to plan about a certain place, he needs to know first the location, socio-economic conditions, means of transport and the required amenities (non-spatial data). He can easily plan about it on his computer screen sitting in his room with the help of remote sensing and GIS.

Taking an example, if a proposal is made to sanction loan/grant-in aid for planting trees in a far away hill district, it can be easily verified whether the place is available or not, without going at that place. This will reduce security problem and save time. Not only this, if a dispensary is to be set up in a hill district, informations regarding the available place for the dispensary, the population of the very area and the common disease that prevail in that place can be easily gathered using remote sensing and GIS.

Therefore, the planners and the head of departments will benefit much from remote sensing, which gives many advantages in every field. For this purpose, a database creation project named Natural resources information system has been taken up by Manipur Remote Sensing Applications Centre under the Department of Space Govt. of India in collaboration with North Eastern Space Applications Centre, Shillong (NSAC). The aim of this project is to have a visual discussion between the decision makers or planners and the farmers at the field to increase the products and to make progress in other development works. The Indian Space Research Organization (ISRO) has established a state Remote Sensing Centre in every state of India. Every Centre has been supplied with a V-SAT. With the help of the V-SAT, a state-wise programme for all the states of India is conducted. The conducted programme is telecast live on the T.V., with the help of this V-SAT.

In that live programme, we can ask questions. For example, a farmer watching that programme can ask which crop would be more suitable to grow in his field, to yield better products. The resource person on the T.V. screen will ask the name and address of the farmer and then, his field will be shown on the screen, i.e., the condition of the soil, area, availability of water, climate condition etc. and after discussing the above factors, he will advice the farmer to grow crops suitable in that place.

Not only this, the concerned department will also get the opportunity to ask questions regarding their progress. Here, we need to recall back by whom and for what purpose, Remote Sensing Technology has been brought into a sub-continent like India. In the early 1960, Sir Vikram Sarabhai, father of space science and Remote Sensing in India had set up this process in India. Regarding why and for what purpose, it is because the population of India kept on increasing and the measures to increase the productivity could not be seen.

It became necessary to find a way of how to use the cultivable land it, area of a place which was not used before for any purpose. Way to find out? To transform into a place of high productivity, but since the country is a large one, to make a survey was not an easy thing. At last, taking photos of the whole surface of the sub-continent was the only option.

Thus, the advantages of remote sensing came to be common. So, till now, the work of Sir Vikram Sarabhai is still remembered by the Remote Sensing Community. With the help of his technology, we can also monitor the various natural disasters. For example, the areas affected by flood, the damages caused by it and the losses it has made can be calculated and the required compensation/relief can be given.

By comparing the flooded areas with the satellite photo of the place before the flood and from the observations made by the concerned scientists, the damages can be easily found out. By estimating the loss or damages from a helicopter, we cannot know the exact loss. Besides, now we have using powerful sensors that, the forest department can even count how many trees are planted at which place and when and how many tress are growing.

In our state also, before, there were only two computers and two scientists in the Remote Sensing Centre, but now the number of computers with powerful software have increased.

There are also five scientists according to their five fields.

They are:-

(i) Soil and Agriculture.

(ii) Land use and Human Settlement.

(iii) Environment and Ecology.

(iv) Water Resource, and

(v) Geosciences.

Regarding roadways, landslide prone areas can be detected and more over in hilly regions, new roads can be constructed places apart from the landslide prone areas. In this case, a research and development project entitled, "Landslide Hazard Zonation" from Imphal to Mao has been undertaken by the Centre under the department of Science and Technology, Govt. of India. Not only this, students, research scholars, environmentalists, local NGOs, etc. have been contributing various required datas and informations to the Remote Sensing Centre.

Achievements in Remote Sensing Applications in India

The pressure of growing population, increased demand for food, fodder and fuelwood combined with rapid industrialization and urbanization have caused tremendous pressure on natural resources and environment. The consequent impact of these activities have given rise to crisis in agreoecosystems like land degradation, accelerated erosion, shifting cultivation, degradation of forests, increase in frequency of floods, pollution, urban sprawl etc. The challenge before us, therefore, is to reverse the process of degradation and undesirable changes in the environment in order to conserve and sustain the resource base of the country to meet the present demand as well as the future needs of the people.

In this context, the direct benefits from space technology and the applications driven approach adopted by the Indian Space Remote Sensing Programme have firmly established its immense potential to cater to the needs of national development. The evolution of National Natural Resource Management System (NNRMS) towards fully harnessing the potentials of space remote sensing and the development of the series of Indian Remote Sensing Satellites, besides establishment of necessary ground based data establishment of necessary ground based data reception processing and dissemination systems as well as remote sensing facilities at National Remote Sensing Agency (NRSA), Space Applications Centre (SAC) and Regional Remote Sensing (Service Centres (RRSSCs) for efficient and effective analysis of remotely sensed data are the major steps accomplished in pursuit of this goal. With the establishment of Remote Sensing Applications Centres in several States under many Governmental organizations, remote sensing today has come to stay as an integral part of the national development efforts in the vital sectors of agriculture, hydrology, geology, forestry, oceanography, mineral resources and distaster management like drought, flood, cyclone, earthquake, landslides crop pests, forest fires etc., thus touching every facet of national development. Today, India has acquired a strong self reliant base to harness the full potential of this technology and as a result, the national objective of achieving sustainable development at microlevel is being addressed through the integration of remotely sensed data with other relevant collateral information to arrive at locale specific, environment friendly, economically viable and culturally acceptable treatment packages.

Space and Ground Segments

The Indian Remote Sensing Satellite (IRS-A), the first in the operational series of Indian Remote Sensing Satellites launched in March 1988, has been functioning satisfactorily for the past 6 ½ year but still continues tpo provide quality data. The second satellite in the IRS series, IRS-IB, launched in August 1991, is performing very well. These two satellites have become the mainstay of the National Natural Resources Management System (NNRMS) IRS-1A/1B provides imagery from the two cameras, linear Imaging Self Scanners, LISS-I with both the resolution 72.5m and LISS-IIA and LISS-IIB both with a resolution of 36.25m. IRS-1A and 1B together provide imagery with a combined repetivity of 11 days. While IRS-IA has so far provided more than 5,00,000 scenes, IRS-1B has provided more than 2,00,000 scenes. The successful implementation of the operational IRS-IA 1B system in the country and the need to cater tot eh enhanced user demands in the 1990's has given rise to configuring the second generation remote sensing satellitesm IRS-IC and ID taking in to account the technology development scenario and user requirements during the nineties.

Future Satellite Programmes

The second generation of IRS Satellites namely IRs-1C and 1D which are now under development for launch in 1995 and 1998 respectively have made very good progress during the year. These satellites will have better spectral and spatial resolutions, more frequent revisits, stereo-viewing and onboard data recording facility. The imaging sensors planned for IRS-IC and ID are a multi-spectral Linear Imaging Self Scanner (LISS-3) in visible and near Ir bands, a short Wave infrared (SWIR) band, a panchromatic camera with stereo viewing capability and a Wide Field Sensor (WIFS) in visible and near IR. It is also planned to have onboard data recording capabilities to record the pay load data for wider data coverage/distribution. IRS-ID identical to IRS-IC will be in- orbit to ensure data continuity beyond the mid nineties. The development flights of PSLV during the timeframe of 1994-96 are progressing well. The IRS-P2 will carry the payload flown onboard IRS-IA and IB in order to provide continuity of data services to the remote sensing user community in the country. The IRS-P3 will carry a payload mix of sensors for application related to oceanography and vegetation dynamics besides a scientific payload is planned to be included of experimental studies in X-ray astronomy. Thus IRS-P3 will carry a Modular Optoelectronic canner (MOS), a Wide field Sensor (WiFS) ad an X-ray astronomy payload.

Table: *Orbital Characteristics of IRS-1C/1D*

Orbit	***Polar sun-synchronous***
Altitude	817 km
Repetitivity	24 days
Local time of Equator Crossing	10.30 AM
Mission Life (planned)	3 years

Data Reception, Processing and Dissemination

The National Remote Sensing Agency (NRSA), Hyderabad continued to acquire and archive data from Indian Remote Sensing Satellites, IRS-IA and IRS-IB. In addition, data from other contemporary satellites, viz, Landsat NOAA and ERS-1 SAR are also being received. Augmention of ground segment elements for IRS-1C/1D data reception and processing is nearing completion. It is proposed to received SeaWiFS data after the launch SEASTAR satellite. Satellite data products are being disseminated to the users in the form of photographic films, paper prints, CCT's floppy and cartridge products. About 705 of the user demands in the country is met by IRs-1A/1B data. As part of aerial remote sensing programme, about 759 ours of aerial flying were undertake during the year mainly for aeromagnetic survey, town and country planning, sea level rise, oil exploration and metropolitan developments.

Integrated Mission for Sustainable Development

Sustainable development of natural resources relies on maintaining the fragile balance between productivity functions and conservation practices through precise identification and systematic monitoring of problem area in various resources and developmental sectors. It also calls for application of alternate agricultural practices, crop rotations use bio-fertilisers, energy-efficient farming methods and reclamation of under utilised and wastelands, planned exploitation of mineral and ground water resources etc.

There fore optimal and exploitation of the resources (both renewable and non-renewable resources) calls for an holistic approach. The synoptic view provide by satellite remote sensing offers technologically the appropriate method for studying land and water resources, characterizing the coherent agricultural zones and identifying the constraints/ecological problems at microlevel. Effective use of space based remote sensing data suitably merged with collateral, socio economic and meteorological data through the use of geographic information system helps arriving at locale specific prescriptions to achieve sustainable development of natural resources of the region. The process of sustainable development packages evolving from such an integrated effort will be economically acceptable and environmentally friendly. Towards this, Department of space in 1988had initiated pilot studies in 21 perennially drought affected districts of the country to generate locale specific action plans. These pilot studies carried out has demonstrated the efficacy of using remote sensing based approach for optimal utilization of land and water resources towards combating drought on a long term basis.

Based on the results obtained form these pilot studies, Department of Space has launched a major, unique programme on Integrated Mission for Sustainable Development. This mission, as of now, covers a total of 157 problem districts of the country, covering nearly 45% of India's geographical area which are perennially affected and tribal areas.

The methodology of study involves generation of thematic maps showing current landuse/landcover, types of wasteland, forest cover/ types, surface water resources. Drainage pattern, potential ground water zones, land forms (geomorphology, geology (rock types, structural features, mineral occurence, soil types etc using data from Indian Remote Sensing Satellite. The map showing slope/aspect is prepared using topographic contour information and the meteorological data (rainfall intensity distribution etc. are collected form existing database.

The integration of the various thematic maps and attributes data, and further analysis for identified alternatives for development are carried out using geographic information system to identify set of coherent micro-level land units which are unique in terms of their resources potential and problems. Specific developmental plans for these units are arrived at in consultation with and close coordination between space scientists, experts from various central/state developmental departments, agricultural universities/research institutions, district level officials and local farmers, so as to ensure the technical feasibility and cultural acceptability of such action plans. Further, methodology has been developed to overlay the land capability map over the cadastral maps to provide information at the field levels. Various elements of such action plans are prioritized and categorised as those relating to land and water resources and others. Sensitivity of each element of the action plan would also be studied to see visible regress over the coming years, once they are implemented. The parameters such as, vegetation index, forest cover, land use, agricultural production, ground water table, soil erosion, etc., would be identified for bench marking the impact of such an implementation.

So far, action plans have been generated for one watershed each in 7 districts. The action plans are being implemented by the district authorities for these watersheds. The initial feedback on the implementation of action plans in the watersheds of Anantapur has been encouraging as shown by increased ground water levels through water harvesting structures and consequem agncumetna activities. The supports for implementation is derived from ongoing developmental programmes such as Drought Prone Area Development Programme (DPAP), Desert Development Programme (DDP), Integrated Wasteland Development Programme (IWDP), National Watersheds Development Project for Rain-fed Areas (NWDPRA), Hill Area Development Programme (HADP), etc. It is also proposed to use satellite data for monitoring of the impact of these action plans.

Further, in view of the advantages of the integrated approach wherein local specific action plans will be generated with cadastral overlays, Ministry of Rural Development has suggested taking up of 92 selected blocks from among the IMSD districts, on an urgent basis in a timeframe of 18 months. Ministry of Rural Development will provide funds for implementation of these action plans through pooling of resources from the various ongoing schemes for rural development.

National (Natural) Resources Information System (NRIS)

A National (Natural Resources Information System (NRIS) for the country has been evolved to aid decision makers at national/regiona/ state/district/taluk levels. A numbe of pilot scale studies on various themes viz., wasteland development, land capability classes, district level planning, regional mineral targeting, etc., were carried out towards implementation of NRIS. Several geographic information packages have also been developed indigenously to cater to computer based decision support systems.

Infrastructure/Facilities Development

Fuller utilisation of this technology by uses has been ensured through establishing appropriate remote sensing facilities at central and state level user departments/agencies. As many as 22 States have established Remote Sensing Application Centres, which are carrying out several application projects of relevance to their region, besides participating in national level projects. They have established a strong interaction with all the concerned users. The DOS Centres such as NRSA, Hyderabad; SAC, Ahmedabad and RRSSCs located at Bangalore, Dehradun, Jodhpur, Kharagpur and Nagpur continued to execute various national level missions/projects, user projects, application validation projects, technology and software development activities and conduct of regular training programmes for users.

Training and Education in Remote Sensing

With the increased awareness and utilisation of remote sensing in India, need for adequate number of trained manpower for harnessing the benefits of remote sensing technology has steeply gone up. Adequate efforts are being made to increase the present throughput of trained manpower (of about 800) in various themes in collaboration with central/state/academic institution by way of augmenting the existing training infrastructure. Regular training courses are being conducted by several training institutions such as Indian Institute of Remote Sensing (IIRS, Dehradun), Centre of Studies for Resources Engineering (IIT, Bombay), Institute of Remote Sensing (Anna University, Madras), Survey of India, ecological Survey of India Training Institute, Forest Survey of India and Indian Agricultural Research Institute (IARI, Delhi). Besides these regular courses, on the job training is being provided by National Remote Sensing Agency (NRSA, Hyderabad) and Space Applications Centre (SAC, Ahmedabad) for specific themes and training in digital analysis and GIS by RRSSCs. So far as many as 6300 scientists have been trained in remote sensing. Efforts are also on to introduce remote sensing evening school and college curriculum. Many universities have introduced remote sensing in their post-graduate courses as a full-fledged degree course or a part of a core subject.

International Collaboration

India has many collaborative programmes with several countries in promoting active cooperation in remote sensing. For example, the ESCAP/UNDP Regional Remote Sensing Programme (RRSP), under the execution of the ESCAP, is playing a crucial rle in promoting active cooperation in remote sensing among member counties of Asia-Pacific region, by bringing together experts from different areas who can share their experiences, disseminating information on available expertise, disseminating information on available expertise, disseminating information on available experts where needed.

The SHARES (Sharing of Experience) program initiated by India is also continuing to provide assistance to candidates from developing countries for participation in remote sensing training courses in India. The data reception/processing courses in India. The data reception/ processing facility at Norman, USA has been recently established, to receive data from India Remote Sensing Satellite (IRS) as part of arrangements between EOSAT, USA and Department of Space/Antrix/ NRSA, India, IRS-IC data is also planned to be acquired at this station once the satellite is launched. The DLR of Germany is providing a payload called Modular Opto-electronic Scanner (MOS) proposed to be flown in IRS-P3.

Chapter 3

Geocoding and Remote Sensing

Geocoding is the process of finding associated geographic coordinates (often expressed as latitude and longitude) from other geographic data, such as street addresses, or zip codes (postal codes). With geographic coordinates the features can be mapped and entered into Geographic Information Systems, or the coordinates can be embedded into media such as digital photographs via geotagging.

Reverse geocoding is the opposite: finding an associated textual location such as a street address, from geographic coordinates.

A geocoder is a piece of software or a (web) service that helps in this process.

Address Interpolation

A simple method of geocoding is address interpolation. This method makes use of data from a street geographic information system where the street network is already mapped within the geographic coordinate space. Each street segment is attributed with address ranges (e.g. house numbers from one segment to the next). Geocoding takes an address, matches it to a street and specific segment (such as a block, in towns that use the "block" convention). Geocoding then interpolates the position of the address, within the range along the segment.

Example

Let's say that this segment (for instance, a block) of Evergreen Terrace runs from 700 to 799. Even-numbered addresses would fall on one side (e.g. west side) of Evergreen Terrace, with odd-numbered addresses on the other side (e.g. east side). 742 Evergreen Terrace would (probably) be located slightly less than halfway up the block, on the west side of the street. A point would be mapped at that location along the street, perhaps offset some distance to the west of the street centerline.

Complicating Factors

However, this process is not always as straightforward as in this example.

Difficulties arise when:

- distinguishing between ambiguous addresses such as 742 Evergreen Terrace and 742 W Evergreen Terrace
- attempting to geocode new addresses for a street that is not yet added to the geographic information system database.

While there might be 742 Evergreen Terrace in Springfield, there might also be a 742 Evergreen Terrace in Shelbyville. Asking for the city name (and state, province, country, etc. as needed) can solve this problem. Some situations require use of postal codes or district name for disambiguation. For example, there are multiple 100 Washington Streets in Boston, Massachusetts because several cities have been annexed without changing street names.

Finally, several caveats on using interpolation:

- The typical attribution of a street segment assumes that all "even" numbered parcels are on one side of the segment, and all "odd" numbered parcels are on the other. This is often not true in real life.
- Interpolation assumes that the given parcels are evenly distributed along the length of the segment. This is almost never true in real life; it is not uncommon for a geocoded address to be off by several thousand feet.
- Segment Information (esp. from sources such as TIGER) includes a maximum upper bound for addresses and is interpolated as though the full address range is used. For example, a segment (block) might have a listed range of 100-199, but the last address at the end of the block is 110. In this case, address 110 would be geocoded to 10% of the distance down the segment rather than near the end.
- Most interpolation implementations will produce a point as their resulting "address" location. In reality, the physical address is distributed along the length of the segment, i.e. consider geocoding the address of a shopping mall- the physical lot may run quite some distance along the street segment (or could be

thought of as a two-dimensional space-filling polygon which may front on several different streets- or worse, for cities with multi-level streets, a three-dimensional shape that meets different streets at several different levels) but the interpolation treats it as a singularity.

A very common error is to believe the accuracy ratings of a given map's geocodable attributes. Such "accuracy" currently touted by most vendors has no bearing on an address being attributed to the correct segment, being attributed to the correct "side" of the segment, nor resulting in an accurate position along that correct segment. With the geocoding process used for U.S. Census TIGER datasets, 5-7.5% of the addresses may be allocated to a different census tract, while 50% of the geocoded points might be located to a different property parcel. Because of this, it is quite important to avoid using interpolated results except for non-critical applications, such as pizza delivery. Interpolated geocoding is usually not appropriate for making authoritative decisions, for example if life safety will be impacted by that decision. Emergency services, for example, do not make an authoritative decision based on their interpolations; an ambulance or fire truck will always be dispatched regardless of what the map says.

Other Techniques

Other means of geocoding might include locating a point at the centroid (centre) of a land parcel, if parcel (property) data is available in the geographic information system database. In rural areas or other places lacking high quality street network data and addressing, GPS is useful for mapping a location. For traffic accidents, geocoding to a street intersection or midpoint along a street centerline is a suitable technique. Most highways in developed countries have mile markers to aid in emergency response, maintenance, and navigation. It is also possible to use a combination of these geocoding techniques- using a particular technique for certain cases and situations and other techniques for other cases.

Uses

Geocoded locations are useful in many GIS analysis and cartography tasks.

Geocoding is common on the web, for services like finding driving directions to or from some address, or finding a list of the geographically nearest store or service locations. Geocoding is one of several methods of obtaining geographic coordinates for geotagging media, such as photographs or RSS items.

Privacy Concerns

The proliferation and ease of access to geocoding (and reverse-geocoding) services raises privacy concerns. For example, in mapping crime incidents, law enforcement agencies aim to balance the privacy rights of victims and offenders, with the public's right to know. Law enforcement agencies have experimented with alternative geocoding techniques that allow them to mask some of the locational detail (e.g., address specifics that would lead to identifying a victim or offender). As well, in providing online crime mapping to the public, they also place disclaimers regarding the locational accuracy of points on the map, acknowledging these location masking techniques, and impose terms of use for the information.

List of some Geocoding Systems

Web services:

- Google Maps Free up to 50,000 queries per day, but with numerous restrictions such as an obligation to display Google Maps pictures when using the service.
- Yahoo PlaceFinder Free up to 50,000 queries per day, but with numerous restrictions such as an obligation to display Yahoo Maps pictures when using the service. Does not include Australia or many Asian countries.
- Bing Maps (Microsoft) Free for "public-facing, non-password protected Web sites"; various commercial licence options.
- OpenStreetMaps Free. Poor coverage as at July 2010.
- USC Geocoder Free in batches of 2,500. US only. Uses various reference data sources.
- MapLarge Geocoder Up to 2,000 free per day in small batches via CSV upload. US only. Corrects spelling and other common errors. Outputs detailed match quality statistics.
- GeocoderUS An open source U.S. Address Geocoder. Unrestricted use for non-profits, and commercial services for a fee.

Offline geocoders:

- Free text geocoder
- Very simple geocoding for OpenStreetMap data.

Other systems (some of these code systems are free for use, others have different licences):

- ISO 6709 Standard Representation for Geographic Point Location by Coordinates

- C-squares- compact encoding of geographic coordinate bounds (latitude-longitude)
- FIPS country codes (FIPS 10-4), area code, administrative, free
- FIPS place codes (FIPS 55) U.S. only, free
- FIPS county codes (FIPS 6-4) US only, free
- FIPS state codes (FIPS 5-2) US only, free
- Canadian Location Code, encodes a weather forecast region
- Geohash, compact string encoding of a geographic coordinate with arbitrary precision, in public domain
- Georef, a military/air naviation coordinate system for point and area identification
- HASC (Hierarchical administrative subdivision codes)
- IATA airport codes, area/point codes, airports
- ICAO airport codes, area/point codes, airports
- IANA country codes similar to ISO 3166-1 alpha-2
- IOC country codes, area, worldwide
- ISO 3166 country and subdivision codes
- ITU-R country codes
- ITU-T country calling codes
- ITU-T mobile calling codes
- Maidenhead Locator System
- MapDot Protocol: world locations coded into a zone sequence, free
- MARC country codes
- Marsden Squares
- NAC, area codes (area can be indefinitely small)
- NUTS area code, partially administrative, worldwide: countries, Europe : country to community
- ONS code, UK only, administrative
- Postal codes, area, worldwide, country-codes by UPU, free
- Quarter Degree Grid Cells
- UN M.49 region codes, area code, continents, countries (like ISO 3166-1 numeric)
- SALB (Second Administrative Level Boundaries), by UN
- SGC codes, Canada only, statistical

- UN/LOCODE, area, administrative, cities
- UTM
- WMO squares.

Reverse Geocoding

Reverse geocoding is the process of returning an estimated street address number as it relates to a given coordinate. For example, a user can click on a road centerline theme (thus providing a coordinate) and have information returned that reflects the estimated house number. This house number is interpolated from a range assigned to that road segment. If the user clicks at the midpoint of a segment that starts with address 1 and ends with 100, the returned value will be somewhere near 50. Note that reverse geocoding does not return actual addresses, only estimates of what should be there based on the predetermined range.

Data Output and Cartography

Cartography is the design and production of maps, or visual representations of spatial data. The vast majority of modern cartography is done with the help of computers, usually using a GIS but production quality cartography is also achieved by importing layers into a design program to refine it. Most GIS software gives the user substantial control over the appearance of the data.

Cartographic work serves two major functions:

> *First, it produces graphics on the screen or on paper that convey the results of analysis to the people who make decisions about resources. Wall maps and other graphics can be generated, allowing the viewer to visualize and thereby understand the results of analyses or simulations of potential events. Web Map Servers facilitate distribution of generated maps through web browsers using various implementations of web-based application programming interfaces (AJAX, Java, Flash, etc.).*

Second, other database information can be generated for further analysis or use. An example would be a list of all addresses within one mile (1.6 km) of a toxic spill.

Graphic Display Techniques

Traditional maps are abstractions of the real world, a sampling of important elements portrayed on a sheet of paper with symbols to represent physical objects. People who use maps must interpret these symbols. Topographic maps show the shape of land surface with contour lines or with shaded relief.

Today, graphic display techniques such as shading based on altitude in a GIS can make relationships among map elements visible, heightening one's ability to extract and analyse information. For example, two types of data were combined in a GIS to produce a perspective view of a portion of San Mateo County, California.

- The digital elevation model, consisting of surface elevations recorded on a 30-meter horizontal grid, shows high elevations as white and low elevation as black.
- The accompanying Landsat Thematic Mapper image shows a false-colour infrared image looking down at the same area in 30-meter pixels, or picture elements, for the same coordinate points, pixel by pixel, as the elevation information.

A GIS was used to register and combine the two images to render the three-dimensional perspective view looking down the San Andreas Fault, using the Thematic Mapper image pixels, but shaded using the elevation of the landforms. The GIS display depends on the viewing point of the observer and time of day of the display, to properly render the shadows created by the sun's rays at that latitude, longitude, and time of day. An archeochrome is a new way of displaying spatial data. It is a thematic on a 3D map that is applied to a specific building or a part of a building. It is suited to the visual display of heat loss data.

Spatial ETL

Spatial ETL tools provide the data processing functionality of traditional Extract, Transform, Load (ETL) software, but with a primary focus on the ability to manage spatial data. They provide GIS users with the ability to translate data between different standards and proprietary formats, whilst geometrically transforming the data en-route.

GIS Developments

Many disciplines can benefit from GIS technology. An active GIS market has resulted in lower costs and continual improvements in the hardware and software components of GIS. These developments will, in turn, result in a much wider use of the technology throughout science, government, business, and industry, with applications including real estate, public health, crime mapping, national defence, sustainable development, natural resources, landscape architecture, archaeology, regional and community planning, transportation and logistics. GIS is also diverging into location-based services (LBS). LBS allows GPS enabled mobile devices to display their location in relation to fixed assets (nearest restaurant, gas station, fire hydrant), mobile assets (friends, children, police car) or to relay their position back to a central server for display or other processing. These services continue to develop with the increased integration of GPS functionality with increasingly powerful mobile electronics (cell phones, PDAs, laptops).

OGC Standards

The Open Geospatial Consortium (OGC) is an international industry consortium of 384 companies, government agencies, universities and individuals participating in a consensus process to develop publicly available geoprocessing specifications. Open interfaces and protocols defined by OpenGIS Specifications support interoperable solutions that "geo-enable" the Web, wireless and location-based services, and mainstream IT, and empower technology developers to make complex spatial information and services accessible and useful with all kinds of applications. Open Geospatial Consortium (OGC) protocols include Web Map Service (WMS) and Web Feature Service (WFS).

GIS products are broken down by the OGC into two categories, based on how completely and accurately the software follows the OGC specifications.

Compliant Products are software products that comply to OGC's OpenGIS Specifications. When a product has been tested and certified as compliant through the OGC Testing Program, the product is automatically registered as "compliant" on this site.

Implementing Products are software products that implement OpenGIS Specifications but have not yet passed a compliance test. Compliance tests are not available for all specifications. Developers can register their products as implementing draft or approved specifications, though OGC reserves the right to review and verify each entry.

Web Mapping

Web mapping is the process of designing, implementing, generating and delivering maps on the World Wide Web and its product. While web mapping primarily deals with technological issues, web cartography additionally studies theoretic aspects: the use of web maps, the evaluation and optimization of techniques and workflows, the usability of web maps, social aspects, and more. Web GIS is similar to web mapping but with an emphasis on analysis, processing of project specific geodata and exploratory aspects. Often the terms web GIS and web mapping are used synonymously, even if they don't mean exactly the same.

In fact, the border between web maps and web GIS is blurry. Web maps are often a presentation media in web GIS and web maps are increasingly gaining analytical capabilities. A special case of web maps are mobile maps, displayed on mobile computing devices, such as mobile phones, smart phones, PDAs, GPS and other devices. If the maps on these devices are displayed by a mobile web browser or web user agent, they can be regarded as mobile web maps. If the mobile web maps also display context and location sensitive information, such as points of interest, the term Location-based services is frequently used."

"The use of the web as a dissemination medium for maps can be regarded as a major advancement in cartography and opens many new opportunities, such as realtime maps, cheaper dissemination, more frequent and cheaper updates of data and software, personalized map content, distributed data sources and sharing of geographic information. It also implicates many challenges due to technical restrictions (low display resolution and limited bandwidth, in particular with mobile computing devices, many of which are physically small, and use slow wireless Internet connections), copyright and security issues, reliability issues and technical complexity. While the first web maps were primarily static, due to technical restrictions, today's web maps can be fully interactive and integrate multiple media. This means that both web mapping and web cartography also have to deal with interactivity, usability and multimedia issues."

Development and Implementation

The advent of web mapping can be regarded as a major new trend in cartography. Previously, cartography was restricted to a few companies, institutes and mapping agencies, requiring expensive and complex hardware and software as well as skilled cartographers and geomatics engineers.

With web mapping, freely available mapping technologies and geodata potentially allow every skilled person to produce web maps, with expensive geodata and technical complexity (data harmonization, missing standards) being two of the remaining barriers preventing web mapping from fully going mainstream. The cheap and easy transfer of geodata across the internet allows the integration of distributed data sources, opening opportunities that go beyond the possibilities of disjoint data storage. Everyone with minimal knowhow and infrastructure can become a geodata provider.

These facts can be regarded both as an advantage and a disadvantage. While it allows everyone to produce maps and considerably enlarges the audience, it also puts geodata in the hands of untrained people who potentially violate cartographic and geographic principles and introduce flaws during the preparation, analysis and presentation of geographic and cartographic data. Educating the general public about geographic analysis and cartographic methods and principles should therefore be a priority to the cartography community.

Types of Web Maps

A first classification of web maps has been made by Kraak. He distinguished *static* and *dynamic* web maps and further distinguished *interactive* and *view only* web maps. However, today in the light of an increased number of different web map types, this classification needs some revision. Today, there are additional possibilities regarding distributed data sources, collaborative maps, personalized maps, etc.

Analytic Web Maps

These web maps offer GIS analysis, either with geodata provided, or with geodata uploaded by the map user. As already mentioned, the borderline between analytic web maps and web GIS is blurry. Often, parts of the analysis are carried out by a serverside GIS and the client displays the result of the analysis. As web clients gain more and more capabilities, this task sharing may gradually shift.

Animated Web Maps

Animated Maps show changes in the map over time by animating one of the graphical or temporal variables. Various data and multimedia formats and technologies allow the display of animated web maps: SVG, Adobe Flash, Java, Quicktime, etc., also with varying degrees of interaction. Examples for animated web maps are weather maps, maps displaying dynamic natural or other phenomena (such as water currents, wind patterns, traffic flow, trade flow, communication patterns,social studies projects, and for college life, etc.).

Collaborative Mapping

Collaborative mapping is the aggregation of web maps and user-generated content, from a group of individuals or entities, and can take several distinct forms.

Types

Collaborative Mapping applications vary depending on which feature the collaborative edition takes place: on the map itself (shared surface), or on overlays to the map. A very simple collaborative mapping application would just plot users' locations (Social mapping or geosocial networking) or Wikipedia articles' locations (Placeopedia). Collaborative implies the possibility of edition by several distinct individuals so the term would tend to exclude applications such as wayfaring where the maps are not meant for the general user to modify.

Shared Surface

In this kind of application, the map itself is created collaboratively by sharing a common surface. For example WikiMapia adds user-generated place names and descriptions to locations. Collaborative mapping and specifically surface sharing faces the same problems as revision control, namely concurrent access issues and versioning. In addition to these problems, collaborative maps must deal with the difficult issue of cluttering, due to the geometric constraints inherent in the media. One approach to this problem is using overlays.

Overlays

Overlays group together items on a map, allowing the user of the map to toggle the overlay's visibility and thus all items contained in the overlay. The application uses map tiles from a third-party (for example one of the mapping APIs) and adds its own collaboratively-edited overlays to them, sometimes in a Wiki fashion. If each user's revisions are contained in an overlay, the issue of revision control and cluttering can be mitigated.

Commercial Context

According to Edward Mac Gillavry there is a dichotomy between corporate projects and user-driven projects. With corporate initiatives generally using a one-way information flow from the service provider to the subscriber and user driven projects generally being characterized by a two way information flow.

Several big internet companies launched mapping applications with collaborative features, most importantly Google Maps with the Google Map Maker feature. Although google allows flexible mash up style use of their raster map images, the MapMaker system presents a one-way flow at the level of raw map data, from the community to google (with the exception of some areas where special provision of shapefiles have been granted for humanitarian reasons). Contrast this with the similar non-corporate collaborative mapping system, OpenStreetMap, which allows all raw map data to be downloaded freely and openly via API requests or full "planet" download.

Customisable Web Maps

Web maps in this category are usually more complex web mapping systems that offer APIs for reuse in other people's web pages and products. Example for such a system with an API for reuse are the Open Layers Framework, Yahoo! Maps and Google Maps.

Distributed Web Maps

These are maps created from a distributed data source. The WMS protocol offers a standardised method to access maps on other servers. WMS servers can collect these different sources, reproject the map layers, if necessary, and send them back as a combined image containing all requested map layers. One server may offer a topographic base map, while other servers may offer thematic layers.

Dynamically Created Web Maps

These maps are created on demand each time the user reloads the webpages, often from dynamic data sources, such as databases. The webserver generates the map using a web map server or a self written software. Some applications refer to depictions as hyper maps. One of the example is- Bhoosampada by Indian Space Research Organizations.

Hyper Maps

Any approach offering the planar presentation of a portion of an n-dimensional orthogonal web map structure with the option to chose the axes for depiction from the dimensions.

Interactive Web Maps

Interactivity is one of the major advantages of screen based maps and web maps. It helps to compensate for the disadvantages of screen and web maps. Interactivity helps to explore maps, change map parameters, navigate and interact with the map, reveal additional information, link to other resources, and much more. Technically, it is achieved through the combination of events, scripting and DOM manipulations.

Online Atlases

Atlas projects often went through a renaissance when they made a transition to a web based project. In the past, atlas projects often suffered from expensive map production, small circulation and limited audience. Updates were expensive to produce and took a long time until they hit the public. Many atlas projects, after moving to the web, can now reach a wider audience, produce cheaper, provide a larger number of maps and map types and integrate with and benefit from other web resources. Some atlases even ceased their printed editions after going online, sometimes offering printing on demand features from the online edition. Some atlases (primarily from North America) also offer raw data downloads of the underlying geospatial data sources.

Personalized Web Maps

Personalized web maps allow the map user to apply his own data filtering, selective content and the application of personal styling and map symbolization. The OGC (Open Geospatial Consortium) provides the SLD standard (Styled Layer Description) that may be sent to a WMS server for the application of individual styles. This implies that the content and data structure of the remote WMS server is properly documented.

Realtime Web Maps

Realtime maps show the situation of a phenomenon in close to realtime (only a few seconds or minutes delay). Data is collected by sensors and the maps are generated or updated at regular intervals or immediately on demand. Examples are weather maps, traffic maps or vehicle monitoring systems.

Static Web Maps

Static web pages are *view only* with no animation and interactivity. They are only created once, often manually and infrequently updated. Typical graphics formats for static web maps are PNG, JPEG, GIF, or TIFF (e.g., drg) for raster files, SVG, PDF or SWF for vector files. Often, these maps are scanned paper maps and had not been designed as screen maps. Paper maps have a much higher resolution and information density than typical computer displays of the same physical size, and might be unreadable when displayed on screens at the wrong resolution.

Temporal Web Maps

Any depiction of a portion of an n-dimensional orthogonal web map structure in a planar projection with time as one of the coordinate axes.

Advantages of Web Maps

- Web maps can easily *deliver up to date information.* If maps are generated automatically from databases, they can display information in almost realtime. They don't need to be printed, mastered and distributed. Examples:
  - A map displaying election results, as soon as the election results become available.
  - A map displaying the traffic situation near realtime by using traffic data collected by sensor networks.
  - A map showing the current locations of mass transit vehicles such as buses or trains, allowing patrons to minimize their

waiting time at stops or stations, or be aware of delays in service.

— Weather maps, such as NEXRAD.

- *Software and hardware infrastructure for web maps is cheap.* Web server hardware is cheaply available and many open source tools exist for producing web maps.
- *Product updates can easily be distributed.* Because web maps distribute both logic and data with each request or loading, product updates can happen every time the web user reloads the application. In traditional cartography, when dealing with printed maps or interactive maps distributed on offline media (CD, DVD, etc.), a map update caused serious efforts, triggering a reprint or remastering as well as a redistribution of the media. With web maps, data and product updates are easier, cheaper, and faster, and can occur more often.
- *They work across browsers and operating systems.* If web maps are implemented based on open standards, the underlying operating system and browser do not matter.
- *Web maps can combine distributed data sources.* Using open standards and documented APIs one can integrate (*mash up*) different data sources, if the projection system, map scale and data quality match. The use of centralized data sources removes the burden for individual organizations to maintain copies of the same data sets. The downside is that one has to rely on and trust the external data sources.
- *Web maps allow for personalization.* By using user profiles, personal filters and personal styling and symbolization, users can configure and design their own maps, if the web mapping systems supports personalization. Accessibility issues can be treated in the same way. If users can store their favourite colours and patterns they can avoid colour combinations they can't easily distinguish (e.g. due to colour blindness).
- *Web maps enable collaborative mapping.* Similar to the Wikipedia project, web mapping technologies, such as DHTML/ Ajax, SVG, Java, Adobe Flash, etc. enable distributed data acquisition and collaborative efforts. Examples for such projects are the OpenStreetMap project or the Google Earth community. As with other open projects, quality assurance is very important, however!
- *Web maps support hyperlinking to other information on the web.* Just like any other web page or a wiki, web maps can act like

an index to other information on the web. Any sensitive area in a map, a label text, etc. can provide hyperlinks to additional information. As an example a map showing public transport options can directly link to the corresponding section in the online train time table.

- *It is easy to integrate multimedia in and with web maps.* Current web browsers support the playback of video, audio and animation (SVG, SWF, Quicktime, and other multimedia frameworks).

Disadvantages of Web Maps and Problematic Issues

- *Reliability issues* – the reliability of the internet and web server infrastructure is not yet good enough. Especially if a web map relies on external, distributed data sources, the original author often cannot guarantee the availability of the information.
- *Geodata is expensive* – Unlike in the US, where geodata collected by governmental institutions is usually available for free or cheap, geodata is usually very expensive in Europe or other parts of the world.
- *Bandwidth issues* – Web maps usually need a relatively high bandwidth.
- *Limited screen space* – Like with other screen based maps, web maps have the problem of limited screen space. This is in particular a problem for mobile web maps and location based services where maps have to be displayed in very small screens with resolutions as low as 100×100 pixels. Hopefully, technological advances will help to overcome these limitations.
- *Quality and accuracy issues* – Many web maps are of poor quality, both in symbolization, content and data accuracy.
- *Complex to develop* – Despite the increasing availability of free and commercial tools to create web mapping and web GIS applications, it is still a complex task to create interactive web maps. Many technologies, modules, services and data sources have to be mastered and integrated.
- *Immature development tools* – Compared to the development of standalone applications with integrated development tools, the development and debugging environments of a conglomerate of different web technologies is still awkward and uncomfortable.
- *Copyright issues* – Many people are still reluctant to publish geodata, especially in the light that geodata is expensive in some

parts of the world. They fear copyright infringements of other people using their data without proper requests for permission.

- *Privacy issues* – With detailed information available and the combination of distributed data sources, it is possible to find out and combine a lot of private and personal information of individual persons. Properties and estates of individuals are now accessible through high resolution aerial and satellite images throughout the world to anyone.

History of Web Mapping

This section contains some of the milestones of web mapping, online mapping services and atlases. Because web mapping depends on enabling technologies of the web, this section also includes a few milestones of the web.

- 1989-90: *Birth of the WWW*, WWW invented at CERN for the exchange of research documents.
- 1990-12: *First Web Browser and Web Server*, Tim Berners-Lee wrote first web browser and web server.
- 1991-04: HTTP 0.9 protocol, Initial design of the HTTP protocol for communication between browser and server.
- 1991-06: *ViolaWWW 0.8 Browser*, The first popular web browser. Written for X11 on Unix.
- 1991-08: *WWW project announced in public newsgroup*, This is regarded as the debut date of the Web. Announced in newsgroup alt.hypertext.
- 1992-06: HTTP 1.0 protocol, Version 1.0 of the HTTP protocol. Introduces the POST method and persistent connections.
- 1993-04: *CERN announced web as free*, CERN announced that access to the web will be free for all. The web gained critical mass.
- 1993-06: HTML 1.0. The first version of HTML, published by T. Berners-Lee and Dan Connolly.
- 1993-07: *Xerox PARC Map Viewer*, The first mapserver based on CGI/Perl, allowed reprojection styling and definition of map extent.
- 1994-06: *The National Atlas of Canada*, The first version of the National Atlas of Canada was released. Can be regarded as the first online atlas.
- 1994-10: *Netscape Browser 0.9 (Mosaic)*, The first version of the highly popular browser Netscape Navigator.

- 1995-03: *Java 1.0*, The first public version of Java.
- 1995-11: HTML 2.0, Introduced forms, file upload, internationalization and client-side image maps.
- 1995-12: *Javascript 1.0*, Introduced first script based interactivity.
- 1995: *MapGuide*, First introduced as Argus MapGuide.
- 1996-01: *JDK 1.0*, First version of the Sun JDK.
- 1996-02: *Mapquest*, The first popular online Address Matching and Routing Service with mapping output.
- 1996-06: *MultiMap*, The UK-based MultiMap website launched offering online mapping, routing and location based services. Grew into one of the most popular UK web sites.
- 1996-11: Geomedia WebMap 1.0, First version of Geomedia WebMap, already supports vector graphics through the use of ActiveCGM.
- 1996-fall: *MapGuide*, Autodesk acquired Argus Technologies.and introduced Autodesk MapGuide 2.0.
- 1996-12: *Macromedia Flash 1.0*, First version of the Macromedia Flash plugin.
- 1997-01: HTML 3.2, Introduced tables, applets, script elements, multimedia elements, flowtext around images, etc.
- 1997-03:Norwegian company Mapnet launches application for www.epi.no with active POI layer for real estate listings.
- 1997-06: *US Online National Atlas Initiative*, The USGS received the mandate to coordinate and create the online National Atlas of the United States of America.
- 1997-07: UMN MapServer 1.0, Developed as Part of the NASA ForNet Project. Grew out of the need to deliver remote sensing data across the web for foresters.
- 1997-12: HTML 4.0, Introduced styling with CSS, absolute and relative positioning of elements, frames, object element, etc.
- 1998-06: *Terraserver USA*, A Web Map Service serving aerial images (mainly b+w) and USGS DRGs was released. One of the first popular WMS. This service is a joint effort of USGS, Microsoft and HP.
- 1998-07: UMN MapServer 2.0, Added reprojection support (PROJ.4).
- 1998-08: MapObjects Internet Map Server, ESRI's entry into the web mapping business.

- 1999-03: HTTP 1.1 protocol, Version 1.1 of the HTTP protocol. Introduces the request pipelining for multiple connections between server and client. This version is still in use as of 2007.
- 1999-08: *National Atlas of Canada, 6th edition*, This new version was launched at the ICA 1999 conference in Ottawa. Introduced many new features and topics. Is being improved gradually, since then, and kept up-to-date with technical advancements.
- 2000-02: ArcIMS 3.0, The first public release of ESRI's ArcIMS.
- 2000-06: ESRI Geography Network, ESRI founded Geography Network to distribute data and web map services.
- 2000-06: UMN MapServer 3.0, Developed as part of the NASA TerraSIP Project. This is also the first public, open source release of UMN Mapserver. Added raster support and support for TrueType fonts (FreeType).
- 2000-08: *Flash Player 5*, This introduced ActionScript 1.0 (ECMAScript compatible).
- 2001-06: MapScript 1.0 for UMN MapServer, Adds a lot of flexibility to UMN MapServer solutions.
- 2001-09: SVG 1.0 W3C Recommendation, SVG (Scalable Vector Graphics) 1.0 became a W3C Recommendation.
- 2001-09: *Tirolatlas*, A highly interactive online atlas, the first to be based on the SVG standard.
- 2002-06: UMN MapServer 3.5, Added support for PostGIS and ArcSDE. Version 3.6 adds initial OGC WMS support.
- 2002-07: ArcIMS 4.0, Version 4 of the ArcIMS web map server.
- 2003-01: SVG 1.1 W3C Recommendation, SVG 1.1 became a W3C Recommendation. This introduced the mobile profiles SVG Tiny and SVG Basic.
- 2003-06: *NASA World Wind*, NASA World Wind Released. An open virtual globe that loads data from distributed resources across the internet. Terrain and buildings can be viewed 3 dimensionally. The (XML based) markup language allows users to integrate their own personal content. This virtual globe needs special software and doesn't run in a web browser.
- 2003-07: UMN MapServer 4.0, Adds 24bit raster output support and support for PDF and SWF.
- 2003-09: *Flash Player 7*, This introduced ActionScript 2.0 (ECMAScript 2.0 compatible (improved object orientation)). Also initial Video Playback support.

- 2004-07: OpenStreetMap was founded by Steve Coast. OSM is a web based collaborative project to create a world map under a free license.
- 2005-01: Nikolas Schiller creates the interactive "Inaugural Map" of downtown Washington, DC
- 2005-02: *Google Maps*, The first version of Google Maps. Based on raster tiles organized in a quad tree scheme, data loading done with XMLHttpRequests. This mapping application became highly popular on the web, also because it allowed other people to integrate google map services into their own website.
- 2005-04: UMN MapServer 4.6, Adds support for SVG.
- 2005-06: *Google Earth*, The first version of Google Earth was released building on the virtual globe metaphor. Terrain and buildings can be viewed 3 dimensionally. The KML (XML based) markup language allows users to integrate their own personal content. This virtual globe needs special software and doesn't run in a web browser.
- 2005-11: *Firefox 1.5*, First Firefox release with native SVG support. Supports Scripting but no animation.
- 2006-05: *Wikimapia* Launched
- 2006-06: *Opera 9*, Opera releases version 9 with extensive SVG support (including scripting and animation).
- 2006-08: SVG 1.2 Mobile Candidate Recommendation, This SVG Mobile Profile introduces improved multimedia support and many features required to build online Rich Internet Applications.
- 2009-01 Nokia makes Ovi Maps free on its smartphones.

Web Mapping Technologies

The potential number of technologies to implement web mapping projects is almost infinite. Any programming environment, programming language and serverside framework can be used to implement web mapping projects. In any case, both server and client side technologies have to be used. Following is a list of potential and popular server and client side technologies utilized for web mapping.

Server Side Technologies

- Web server – The webserver is responsible for handling http requests by web browsers and other user agents. In the simplest case they serve static files, such as HTML pages or static image files. Web servers also handle authentication, content negotiation, server side includes, URL rewriting and forward requests to dynamic resources, such as CGI applications or serverside scripting languages. The functionality of a webserver can usually be enhanced using modules or extensions. The most popular web server is Apache, followed by Microsoft Internet Information Server and others.

— CGI (common gateway interface) applications are executables running on the webserver under the environment and user permissions of the webserver user. They may be written in any programming language (compiled) or scripting language (e.g. perl). A CGI application implements the common gateway interface protocol, processes the information sent by the client, does whatever the application should do and sends the result back in a web-readable form to the client. As an example a web browser may send a request to a CGI application for getting a web map with a certain map extent, styling and map layer combination. The result is an image format, e.g. JPEG, PNG or SVG. For performance enhancements one can also install CGI applications such as FastCGI. This loads the application after the web server is started and keeps the application in memory, eliminating the need to spawn a separate process each time a request is being made.

— Alternatively, one can use scripting languages built into the webserver as a module, such as PHP, Perl, Python, ASP, Ruby, etc. If built into the web server as a module, the scripting engine is already loaded and doesn't have to be loaded each time a request is being made.

- Web application servers are middleware which connects various software components with the web server and a programming language. As an example, a web application server can enable the communication between the API of a GIS and the webserver, a spatial database or other proprietary applications. Typical web application servers are written in Java, C, C++, C# or other scripting languages. Web application servers are also useful when developing complex realtime web mapping applications or Web GIS.
- Spatial databases are usually object relational databases enhanced with geographic data types, methods and properties. They are necessary whenever a web mapping application has to deal with dynamic data (that changes frequently) or with

huge amount of geographic data. Spatial databases allow spatial queries, sub selects, reprojections, geometry manipulations and offer various import and export formats. A popular example for an open source spatial database is PostGIS. MySQL also implements some spatial features, although not as mature as PostGIS. Commercial alternatives are Oracle Spatial or spatial extensions of Microsoft SQL Server and IBM DB2. The OGC Simple Features for SQL Specification is a standard geometry data model and operator set for spatial databases. Most spatial databases implement this OGC standard.

- WMS server are specialized web mapping servers implemented as a CGI application, Java Servlet or other web application server. They either work as a standalone web server or in collaboration with existing web servers or web application servers (the general case). WMS Servers can generate maps on request, using parameters, such as map layer order, styling/ symbolization, map extent, data format, projection, etc. The OGC Consortium defined the WMS standard to define the map requests and return data formats. Typical image formats for the map result are PNG, JPEG, GIF or SVG. There are open source WMS Servers such as UMN Mapserver and Mapnik. Commercial alternatives exist from most commercial GIS vendors, such as ESRI ArcIMS, ArcGIS Server, GeoClip, Intergraph Geomedia WebMap, and others.

Client side technologies

- Web browser – In the simplest setup, only a web browser is required. All modern web browsers support the display of HTML and raster images (JPEG, PNG and GIF format). Some solutions require additional plugins.
  - ECMAScript support – ECMAScript is the standardized version of JavaScript. It is necessary to implement client side interaction, refactoring of the DOM of a webpage and for doing network requests. ECMAScript is currently part of any modern web browser.
  - Events support – Various events are necessary to implement interactive client side maps. Events can trigger script execution or SMIL operations. We distinguish between:
    - Mouse events (mousedown, mouseup, mouseover, mousemove, click)
    - Keyboard events (keydown, keypress, keyup)

  - State events (load, unload, abort, error)
  - Mutation events (reacts on modifications of the DOM tree, e.g. DOMNodeInserted)
  - SMIL animation events (reacts on different states in SMIL animation, beginEvent, endEvent, repeatEvent)
  - UI events (focusin, focusout, activate)
  - SVG specific events (SVGZoom, SVGScroll, SVGResize)
- Network requests – This is necessary to load additional data and content into a web page. Most modern browsers provide the XMLHttpRequest object which allows for get and post http requests and provides some feedback on the data loading state. The data received can be processed by ECMAScript and can be included into the current DOM tree of the web page/web map. SVG user agents alternatively provide the getURL() and postURL() methods for network requests. It is recommended to test for the existence of a network request method and provide alternatives if one method isn't present. As an example, a wrapper function could handle the network requests and test whether XMLHttpRequests or getURL() or alternative methods are available and choose the best one available. These network requests are also known under the term Ajax.
- DOM support – The Document Object Model provides a language independent API for the manipulation of the document tree of the webpage. It exposes properties of the individual nodes of the document tree, allows to insert new nodes, delete nodes, reorder nodes and change existing nodes. DOM support is included in any modern web browser. DOM support together with scripting is also known as DHTML or Dynamic HTML. Google Maps and many other web mapping sites use a combination of DHTML, Ajax, SVG and VML.
- SVG support or SVG image support – SVG is the abbreviation of "Scalable Vector Graphics" and integrates vector graphics, raster graphics and text. SVG also supports animation, internationalization, interactivity, scripting and XML based extension mechanisms. SVG is a huge step forward when it comes to delivering high quality, interactive maps. At the time of writing (2007–01), SVG is natively supported in Mozilla/Firefox >version 1.5, Opera >version 9 and the developer version of Safari/Webkit. Internet

Explorer users still need the Adobe SVG viewer plugin provided by Adobe.

— Java support – some browsers still provide old versions of the Java virtual machine. An alternative is the use of the Sun Java Plugin. Java is a full featured programming language that can be used to create very sophisticated and interactive web maps. The Java2D and Java3D libraries provide 2d and 3d vector graphics support. The creation of Java based web maps requires a lot of programming know how. Adrian Herzog discusses the use of Java applets for the presentation of interactive choroplethe and cartogram maps.

— Web browser plugins

  - Adobe Acrobat – provides vector graphics and high quality printing support. Allows toggling of map layers, hyper links, multimedia embedding, some basic interactivity and scripting (ECMAScript).
  - Adobe Flash – provides vector graphics, animation and multimedia support. Allows the creation of sophisticated interactive maps, as with Java and SVG. Features a programming language (ActionScript) which is similar to ECMAScript. Supports Audio and Video.
  - Apple Quicktime – Adds support for additional image formats, video, audio and Quicktime VR (Panorama Images). Only available to Mac OS X and Windows.
  - Adobe SVG viewer – provide SVG 1.0 support for web browsers, only required for Internet Explorer Users, because it doesn't yet natively support SVG. The Adobe SVG viewer isn't developed any further and only fills the gap until Internet Explorer gains native SVG support.
  - Sun Java plugin provides support for newer and advanced Java Features.

Global Change, Climate History Program and Prediction of its Impact

Maps have traditionally been used to explore the Earth and to exploit its resources. GIS technology, as an expansion of cartographic science, has enhanced the efficiency and analytic power of traditional mapping. Now, as the scientific community recognizes the environmental consequences of anthropogenic activities influencing climate change, GIS technology is becoming an essential tool to understand the impacts of this change over time. GIS enables the combination of various sources of data with existing maps and up-to-date information from earth observation satellites along with the outputs of climate change models. This can help in understanding the effects of climate change on complex natural systems. One of the classic examples of this is the study of Arctic Ice Melting. The outputs from a GIS in the form of maps combined with satellite imagery allow researchers to view their subjects in ways that literally never have been seen before. The images are also invaluable for conveying the effects of climate change to non-scientists.

Adding the Dimension of Time

The condition of the Earth's surface, atmosphere, and subsurface can be examined by feeding satellite data into a GIS. GIS technology gives researchers the ability to examine the variations in Earth processes over days, months, and years.

As an example, the changes in vegetation vigor through a growing season can be animated to determine when drought was most extensive in a particular region. The resulting graphic, known as a normalized vegetation index, represents a rough measure of plant health. Working with two variables over time would then allow researchers to detect regional differences in the lag between a decline in rainfall and its effect on vegetation.

GIS technology and the availability of digital data on regional and global scales enable such analyses. The satellite sensor output used to generate a vegetation graphic is produced for example by the Advanced Very High Resolution Radiometer (AVHRR). This sensor system detects the amounts of energy reflected from the Earth's surface across various bands of the spectrum for surface areas of about 1 square kilometer. The satellite sensor produces images of a particular location on the Earth twice a day. AVHRR and more recently the Moderate-Resolution Imaging Spectroradiometer (MODIS) are only two of many sensor systems used for Earth surface analysis. More sensors will follow, generating ever greater amounts of data.

GIS and related technology will help greatly in the management and analysis of these large volumes of data, allowing for better understanding of terrestrial processes and better management of human activities to maintain world economic vitality and environmental quality.

In addition to the integration of time in environmental studies, GIS is also being explored for its ability to track and model the progress of humans throughout their daily routines. A concrete example of progress in this area is the recent release of time-specific population data by the US Census. In this data set, the populations of cities are shown for daytime and evening hours highlighting the pattern of concentration and dispersion generated by North American commuting patterns. The manipulation and generation of data required to produce this data would not have been possible without GIS.

Using models to project the data held by a GIS forward in time have enabled planners to test policy decisions. These systems are known as Spatial Decision Support Systems.

Semantics

Tools and technologies emerging from the W3C's Semantic Web Activity are proving useful for data integration problems in information systems. Correspondingly, such technologies have been proposed as a means to facilitate interoperability and data reuse among GIS applications and also to enable new analysis mechanisms.

Ontologies are a key component of this semantic approach as they allow a formal, machine-readable specification of the concepts and relationships in a given domain. This in turn allows a GIS to focus on the intended meaning of data rather than its syntax or structure. For example, reasoning that a land cover type classified as *deciduous needleleaf trees* in one dataset is a specialization or subset of land cover type *forest* in another more roughly classified dataset can help a GIS automatically merge the two datasets under the more general land cover classification. Tentative ontologies have been developed in areas related to GIS applications, for example the hydrology ontology developed by the Ordnance Survey in the United Kingdom and the SWEET ontologies developed by NASA's Jet Propulsion Laboratory. Also, simpler ontologies and semantic metadata standards are being proposed by the W3C Geo Incubator Group to represent geospatial data on the web.

Recent research results in this area can be seen in the International Conference on Geospatial Semantics and the Terra Cognita — Directions to the Geospatial Semantic Web workshop at the International Semantic Web Conference.

Neogeography

Neogeography literally means "new geography" (aka Volunteered Geographic Information), and is commonly applied to the usage of geographical techniques and tools used for personal and community activities or for utilization by a non-expert group of users. Application domains of neogeography are typically not formal or analytical.

History

The term neogeography has been used since at least 1922. In the early 1950s in the U.S. it was a term used in the sociology of production & work. The French philosopher Francois Dagognet used it in the title of his 1977 book *Une Epistemologie de l'espace concret: Neo-geographie.* The word was first used in relation to the study of online communities in the 1990s by Kenneth Dowling, the Librarian of the City and County of San Francisco. Immediate precursor terms in the industry press were: "the geospatial Web" and "the geoaware Web" (both 2005); "Where 2.0" (2005); "a dissident cartographic aesthetic" and "mapping and counter-mapping" (2006). These terms arose with the concept of Web 2.0, around the increased public appeal of mapping and geospatial technologies that occurred with the release of such tools as "slippy maps" such as Google Maps, Google Earth, and also with the decreased cost of geolocated mobile devices such as GPS units. Subsequently, the use of geospatial technologies began to see increased integration with non-geographically focused applications.

The term neogeography was first defined in its contemporary sense by Randall Szott on 7 April 2006, and elaborated on May 27, 2006. He argued for a broad scope, to include artists, psychogeography, and more. The technically-oriented aspects of the field, far more tightly defined than in Szott's definition, were outlined by Andrew Turner in his *Introduction to Neogeography* (O'Reilly, 2006). The contemporary use of the term, and the field in general, owes much of its inspiration to the locative media movement that sought to expand the use of location-based technologies to encompass personal expression and society.

Traditional Geographic Information Systems historically have developed tools and techniques targeted towards formal applications that require precision and accuracy. By contrast, neogeography tends to apply to the areas of approachable, colloquial applications. The two realms can have overlap as the same problems are presented to different sets of users: experts and non-experts.

User-generated Geographic Content

Neogeography has also been connected with the increase in user-generated geographic content, closely related to Volunteered Geographic Information. This can be active collection of data such as OpenStreetMap, or passive collection of user-data such as Flickr tags for folksonomic toponyms.

Discussion about the Definition

There is currently much debate about the scope and application of Neogeography in the web mapping, geography, and GIS fields. Some of this discussion considers neogeography to be the ease of use of geographic tools and interfaces while other points focus on the domains of application. Neogeography is not limited to a specific technology and is not strictly web-based, so is not synonymous with web mapping though it is commonly conceived as such.

A number of geographers and geoinformatics scientists (such as Mike Goodchild) have expressed strong reservations about the term "neogeography". They say that geography is an established scientific discipline; uses such as mashups and tags in Google Earth are not scientific works, but are better described as Volunteered Geographic Information. There are also a great many artists and inter-disciplinary practitioners involved in an engagement with new forms of mapping and locative art. It is thus far wider than simply web mapping.

Public Participation GIS

Public Participation Geographic Information Systems (PPGIS) was born, as a term, in 1996 at the meetings of the National Centre for Geographic Information and Analysis (NCGIA). PPGIS is meant to bring the academic practices of GIS and mapping to the local level in order to promote knowledge production. The idea behind PPGIS is empowerment and inclusion of marginalized populations, who have little voice in the public arena, through geographic technology education and participation. PPGIS uses and produces digital maps, satellite imagery, sketch maps, and many other spatial and visual tools, to change geographic involvement and awareness on a local level.

Applications

Attendees to the *Mapping for Change International Conference on Participatory Spatial Information Management and Communication* conferred to at least three potential implications of PPGIS; it can: (1) enhance capacity in generating, managing, and communicating spatial information; (2) stimulate innovation; and ultimately; (3) encourage positive social change. This reflects on the rather nebulous definition of PPGIS as referenced in the *Encyclopedia of GIS* which describes PPGIS as having a definition problem.

There are a range of applications for PPGIS. The potential outcomes can be applied from community and neighbourhood planning and development to environmental and natural resource management.

Marginalized groups, be they grassroots organizations to indigenous populations could benefit from GIS technology.

Governments, non-government organizations and non-profit groups are a big force behind many programs. The current extent of PPGIS programs in the US has been evaluated by Sawicki and Peterman. They catalog over 60 PPGIS programs who aid in "public participation in community decision making by providing local-area data to community groups," in the United States (Craig et al., 2002:24). The organizations providing these programs are mostly universities, local chambers of commerce, non-profit foundations.

In general, neighbourhood empowerment groups can form and gain access to information that is normally very easy for the official government and planning offices to obtain. It is easier for this to happen than for individuals of lower-income neighbourhoods just working by themselves. There have been several projects where university students help implement GIS in neighbourhoods and communities.

It is believed that access to information is the doorway to more effective government for everybody and community empowerment. In a case study of a group in Milwaukee, residents of an inner-city neighbourhood became active participants in building a community information system, learning to access public information and create and analyse new databases derived from their own surveys, all with the purpose of making these residents useful actors in city management and in the formation of public policy. In many cases, there are providers of data for community groups, but the groups may not know that such entities exist. Getting the word out would be beneficial.

Some of the spatial data that the neighbourhood wanted was information on abandoned or boarded-up buildings and homes, vacant lots, and properties that contained garbage, rubbish and debris that contributed to health and safety issues in the area. They also appreciated being able to find landlords that were not keeping up the properties.

The university team and the community were able to build databases and make maps that would help them find these areas and perform the spatial analysis that they needed. Community members learned how to use the computer resources, ArcView 1.0, and build a theme or land use map of the surrounding area. They were able to perform spatial queries and analyse neighbourhood problems. Some of these problems included finding absentee landlords and finding code violations for the buildings on the maps (Ghose 2001).

Approaches

There are two approaches to PPGIS use and application. These two perspectives, top-down and bottom-up, are the currently debated schism in PPGIS.

Top-down

According to Sieber (2006), PPGIS was first envisioned as a means of mapping individuals by many social and economic demographic factors in order to analyse the spatial differences in access to social services. She refers to this kind of PPGIS as *top-down*, being that it is less hands on for the public, but theoretically serves the public by making adjustments for the deficiencies, and improvements in public management.

Bottom-up

A current trend with academic involvement in PPGIS, is researching existing programs, and or starting programs in order to collect data on the effectiveness of PPGIS. Elwood (2006) in *The Professional Geographer*, talks in depth about the "everyday inclusions, exclusions, and contradictions of Participatory GIS research." The research is being conducted in order to evaluate if PPGIS is involving the public equally. In reference to Sieber's top-down PPGIS, this is a counter method of PPGIS, rightly referred to as *bottom-up* PPGIS. Its purpose is to work with the public to let them learn the technologies, then producing their own GIS.

Public Participation GIS is defined by Sieber as the use of geographic information systems to broaden public involvement in policymaking as well as to the value of GIS to promote the goals of nongovernmental organizations, grassroots groups and community based organizations (Sieber 2006). It would seem on the surface that PPGIS, as it is commonly referred to, in this sense would be of a beneficial nature to those in the community or area that is being represented. But in truth only certain groups or individuals will be able to obtain the technology and utilize it. Is PPGIS becoming more available to the underprivileged sector of the community? The question of "who benefits?" should always be asked, and does this harm a community or group of individuals.

The local, participatory management of urban neighbourhoods usually follows on from 'claiming the territory', and has to be made compatible with national or local authority regulations on administering, managing and planning urban territory (McCall 2003). PPGIS applied to participatory Community/Neighbourhood Planning has been examined by, among many others. Specific attention has been given to applications such as housing issues or neighbourhood revitalization. Spatial databases along with the P-mapping are used to maintain a public records GIS or community land information systems. These are just a few of the uses of GIS in the community.

Remote Sensing and GIS Integration in the Work of the Syro-Hungarian Archaeological Mission

The coastal area of the Syrian Arab Republic has been inhabited since the earliest times of history and the intensity of human activity reflected in the settlement pattern must have produced considerable traces during the millenia. Due to the fact that the coast has always been a suitable region for living, there was no lack of people that settled former sites, in many cases reshaping, altering or erasing them.

Due to the rapid development in recent decades, the Syrian coast is undergoing its most thorough transformation during its long history both in the field of population growth plus the accompanying building activity and the change of the natural environment, due to the expansion of agricultural activity. The expansion poses a special threat to the lesser archaeological sites, the visible remains of which are usually hidden in remote areas or lie under the surface of the ground unexplored.

In order to save the unique cultural heritage of the syrian coast, this development should go side by side with the documentation of the archaeological sites, many of which are still waiting to be explored. Since its foundation in 2000, the main activity of the Syro-Hungarian Archaeological Mission in Tartous Governorate is the exploration and the documentation of the archaeological sites dating between the Late Antiquity and the end of the Middle Ages. Special attention is devoted to the study of the rural hinterland of large settlement centres, without which we would only have an incomplete understanding of their working mechanisms. As rural sites (mainly villages) and the elements of the attached economic infrastructure (roads, dams, mills, olive presses...etc.) leave little trace in an intensively used area in such a long timespan, the study of such subject needs a highly integrated approach. This approach is in large part determined by the special caracteristics of the terrain and the nature of the archaeological sites in question.

Beside the traditional sources (historical sources, travellers accounts, study of toponymical data) our surveys make extensive use of the information that is provided by the ever helpful inhabitants of the countryside. The sourvey route planning is foremostly dependant on this data, which then will be supplemented by the detailed archaeological surveys accompanied by the architectural documentation and the collecting and study of the field finds. With the exception of a handful of cases the remote sensing and GIS technologies applied by us hiertho, had their main role in the documentation and presentation part of the work. However as the next chosen examples might highlight it, the ratio between the traditional methods and the new technologies could be more evenly balanced in the future.

Documentation of Medieval Rural Sites

One of the main aims of the Syro-Hungarian Archaeological Mission is to shed more light on the life in the agricultural hinterland of the coastal cities and large settlements. Perhaps the most suitable hitorical period for a sample study is the 12-13th century, when the establishment of the crusader statement the introduction of a feudal system more or less similar to that in Europe, which was caracterized by a larger attention paid to the properties in the countryside.

The Latin documents preserved the names of dozens of villages the identifiction of which in the ground is one of the main aims of our surveys. The more rurally oriented crusader system of administration also resulted in the construction of small towers in the centres of the estates. These solid constructions have a much better survival rate than any other medieval village structures. The site of 'Asur near Safita might be the site of Asor mentioned in a Latin charter of the 13th century. This is strenghtened by the remains of a medieval tower found our field survey of 2002. Due to the dense vegetation, mostly in the form of olive plantations, neither this architectural feature, nor the civilian settlement accompanying it would have been traceable on satellite imaginery. Medieval villages in the coastal region seem to have been built with dry stone walls of mainly rubble stone that after their area has been subjected to intensive ploughing leave little trace above ground. This caracteristic that makes the remote sensing technology hardly useful in exploration, makes it indispensable in documentation. The documentation of the approximate boundaries of a medieval rural settlement is mainly depending on the demarcation of the limits of the omnipresent pottery, scattered on the surface of the site.

The use of GPS is indispensable in the fast demarcation of the extent of the sites and is also useful in recording the position of the other small archaeological features like cisterns and olive grinding stones, which have a better survival rate than other elements of the infrastructure. If many similar sites are organized in a purpose built GIS database, even this relatively scanty data can provide useful base for comparison and general observations.

Retracing Former Structures

Many ancient settlements are continuously inhabited, that in many cases makes the precise documentation of the extent of ancient pottery scatter impossible. In the case of Samka, additional information on the approximate extent of the inhabited area could be gained from the cadasterial maps of the French mandate period. These maps demarcated the positions of the ancient cisterns that provided the water supply of the hilltop village up to the 20th century.

Since most of them went out of use and is hard to find, plus the change in the layout of the settlement in Samke has made their retracing more complicated, the use of a GPS is again makes the work easier. As the surveyors of the beginning of the 20th century were mainly interested in structures still in use, our archaeologically focused interest might result in the discovery of more cisterns that were already out of use at the time of the cadasterial survey and thus excluded from the documentation, but which with the help of the GPS could supplement the former data on the extent of the ancient settlement.

Documenting Large Standing Remains

Our surveys in the past years have found the clear traces of a network of antique and late antique settlements that in almost all aspects could be compared to the much better known „Dead Cities". Being similar to the often grandiose structures of the interior areas of Syria in the Byzantine times, one would expect the satellite imaginary to play a much larger role in the exploration part than with the small rural sites mentioned above. However the remnants of the Dead Cities of the coast are mostly just as well hidden from the eyes of the satellites as the extent of the pottery scatter of the medieval village sites.

It is in part due to the vegetation covering, but in a larger part results from the agricultural activity that in the mountains depends almost exclusively on the terracing. Sometimes large stretches of ancient structures are included in terrace walls that are visible from the sides but are practically invisible from a very high altitude. The situation is further complicated by the fact that many Late Antique settlements seem to have been used afterwards as well, which resulted in their sometimes almost complete obliteration by later structures. In these cases field surveys and architectural documentation on the field is indispensable, but this can be very well supplemented by the use of satellite images, especially when drawing a complete plan of the site with recent periods included. Satellite images play a very useful role at the demarcation of the boundaries of extensive sites and the protected area around them.

3D Modelling of Archaeological Sites

The mountainous hinterland of the Syrian coast has several sites that are best understood if set in 3D terrain models that are included in their documentation. The medieval castle of Qulay'a perched on a high vulcanic rock is one of the most vivid example of the need for a complex approach of documentation. In making a 3D model of the terrain of the archaeological site the satellite imaginery is of important use, but the actual documentation of the hardly visible archaeological remains again has to include a thorough field survey including clearing of the vegetation and the use of a TotalStation measuring equipment. The use of a TotalStation can also help in producing a more detailed model of the terrain.

In the above mentioned cases a certain gap could be perceived between the traditional field survey methods and the most widely used remote sensing methods depending on images taken from high altitude. However this gap could be abridged at most of the site types mentioned, if the taking of low altitude images of the archaeological sites were introduced systematically. Amongst the remote sensing technologies applicable for the kind of archaeological survey work carried out by the Syro-Hungarian Archaeological Mission the most significant are aerial photography, satellite imagery, LIDAR, and the different types of geophysical surveys. In the case of the Syrian Arab Republic, the most readily available remote sensing data are the high resolution (0.5-1 m/pixel) satellite images. Other types of imagery are not easy, if not impossible to obtain since they may carry sensitive information.

However, unlike in other, more open areas of Syria, in the Coastal Region the usability of satellite imagery, even that of today's high resolution Quickbird and Ikonos images, is limited. Apart from the few kilometres wide coastal strip practically the whole region is hilly terrain with its larger part being terraced and covered by olive groves or scrub layers. In most cases the rather small archaeological features in the research area of the Syro-Hungarian Mission are hardly if at all detectable in vertical satellite imagery. The above described problems do not mean that satellite imagery has completely been unusable in our work: we have applied rectified space images loaded into our PDA with GPS for navigation on the terrain, and they are also very useful when dealing with open areas or with larger open sites and still upstanding remains to establish a general plan.

However, we think that another approach to remote sensing is needed for a more efficient reconnaissance and survey work in the Coastal Region. As security considerations are understandibly limiting the taking of high altitude aerial photography, we propose low altitude, limited surveys to be carried out in selected areas using tandem motor paraglider aircraft that can take off near or within the surveyed micro-region, can stay in the air only for 1-2 hours, flies at low speed (30-40 km/h), and usually flies at low altitudes, i.e., a few hundred metres above the terrain. Similar survey work has already been done in Armenia by members or the Aerial Archaeological Research Group (researchers R. Palmer and Ch. Musson), and also in Hungary by the co-author of the present paper (Baranya County Museums).

In Baranya County, Hungary, as part of a multinational archaeological survey project co-financed by the Culture 2000 programme of the European Union and numerous co-organizers, we carry out repeated aerial surveys. The aerial survey has a multiple role: site discovery, monitoring of known sites, and aiding field survey and excavation work.

About the Workflow

During the survey work we take low altitude oblique and near-vertical digital aerial photographs of archaeological sites both from small aircraft (such as the Cessna 172) and from paraglider aircraft. Normal survey flights are flown at 500-800 m altitude with a normal aircraft using a Canon 20D digital camera (8 megapixel CCD) 3-5 cm/ pixel and Canon IS USM 17-85 mm zoom lens. The resolution of the rectified images ranges from 20 to 50 cm and with 1-3 m accuracy on the terrain. We have used paragliders for limited area surveys and for recording excavation features. Using the above mentioned camera we could achieve 3-6 cm resolution and an accuracy that often meets the 1:20 scale excavation drawings, but generally remains around or below 10 cm.

Though for a broader area survey a normal, four-seat airplane is more suitable because of its higher speed, it is the low speed and low altitude of the paraglider that provides better quality and higher resolution images that can be orthorectified to a high accuracy using auxiliary software, such as RadCor for the correction of radial lens distortion, and digital terrain model.

In the first case we generate the terrain model by digitizing the contour lines from 1:10.000 maps. In case of small areas and high resolution photography of individual sites the terrain model for the orthorectification is made by using a Total Station (similarly to the aforementioned Qulay'a survey in Syria).

The othorectified images are then used for mapping and delimiting archaeological sites both for our GIS database and for the National

Heritage Database. The orthorectified images also aid our fieldwork: we load both the images and the site maps into a PDA with GPS that we use for field-walking. With the help of the images, finds can be collected even from individual features of an archaeological site. This method in turn enhances the identification, understanding, and analysis of sites without excavation.

Considering the working conditions and the high level of co-operation we experienced in the Syrian Arab Republic we are convinced that a limited, aerial survey campaign focussed on small areas of interest could be done in co-operation with experts from the GORS and the DGAM in the near future. This could significantly enhance our knowledge of the coastal area and thus help in preserving the unique archaeological and cultural heritage of Syria.

Photogrammetry

Photogrammetry is the practice of determining the geometric properties of objects from photographic images. Photogrammetry is as old as modern photography and can be dated to the mid-nineteenth century. In the simplest example, the distance between two points that lie on a plane parallel to the ‘. Photogrammetry is used in different fields, such as topographic mapping, architecture, engineering, manufacturing, quality control, police investigation, angy]], as well as by archaeologists to quickly produce plans of large or complex sites and by meteorologists as a way to determine the actual wind speed of a tornado where objective weather data cannot be obtained. It is also used to combine live action with computer generated imagery in movie post-production; *Fight Club* is a good example of the use of photogrammetry in film.

Algorithms for photogrammetry typically express the problem as that of minimizing the sum of the squares of a set of errors. This minimization is known as bundle adjustment and is often performed using the Levenberg–Marquardt algorithm

Photogrammetry uses methods from many disciplines including optics and projective geometry. The data model on the right shows what type of information can go into and come out of photogrammetric methods.. Each of the four main variables can be an *input* or an *output* of a photogrammetric method.

Photogrammetry has been defined by ASPRS as the art, science, and technology of obtaining reliable information about physical objects and the environment through processes of recording, measuring and interpreting photographic images and patterns of recorded radiant electromagnetic energy and other phenomena. Photogrammetric data with dense range data from scanners complement each other. Photogrammetry is more accurate in the x and y direction while range data is generally more accurate in the z direction.

This range data can be supplied by techniques like LiDAR, Laser Scanners (using time of flight, triangulation or interferometry), White-light digitizers and any other technique that scans an area and returns x, y,z coordinates for multiple discrete points (commonly called "point clouds"). Photos can clearly define the edges of buildings when the point cloud footprint can not. It is beneficial to incorporate the advantages of both systems and integrate it to create a better product. Techniques such as adaptive least squares stereo matching are then used to produce a dense array of correspondences which are transformed through a camera model to produce a dense array of x, y, z data which can be used to produce digital terrain model and orthoimage products.

Systems which use these techniques, e.g the ITG system were developed in the 1980s and 1990s but have since been supplanted by LiDAR and Radar based approaches, although these techniques may still be useful deriving elevation models from old aerial photographs and/or satellite images.

This method is commonly employed in collision engineering, especially with automobiles. When litigation for accidents occur and engineers need to determine the exact deformation present in the vehicle, it is common for several years to have passed and the only evidence that remains are crime scene photographs taken by the police. Photogrammetry is used to determine how much the car in question was deformed, which relates to an amount of energy required to produce that deformation. The energy can then be used to determine important information about the crash (like velocity at time of impact).

Chapter 4

Global Positioning System

Global Positioning System or GPS, is a technology that can give your accurate position anywhere on earth (latitude/longitude). You need a special GPS receiver that can receive signals from satellites. It provides location and time information in all weather, anywhere on or near the Earth, where there is an unobstructed line of sight to four or more GPS satellites. It is maintained by the United States government and is freely accessible by anyone with a GPS receiver with some technical limitations which are only removed for military users. Global Positioning System (GPS) receivers can also determine altitude by trilateration with four or more satellites. In aircraft, altitude determined using autonomous GPS is not precise or accurate enough to supersede the pressure altimeter without using some method of augmentation. In hiking and climbing, it is not uncommon to find that the altitude measured by GPS is off by as much as a thousand meters, if all the available satellites happen to be close to the horizon. The GPS program provides critical capabilities to military, civil and commercial users around the world. It is an engine of economic growth and jobs, and has generated billions of dollars of economic activity. It maintains future warfighter advantage over opponents and is one of the four core military capabilities.

It works anywhere on the planet where you can receive signals from the satellite. You will need a pretty clear view of the sky for GPS to work, so it won't work inside buildings, underground or even in a forest. Few years back, GPS wasn't available to it's full accuracy, but since the year 2000 it is available with full accuracy anywhere in the world and to anyone who buys a GPS receiver. In addition, GPS is the backbone for modernising the global air traffic system. The GPS project was developed in 1973 to overcome the limitations of previous navigation systems, integrating ideas from several predecessors, including a number of classified engineering design studies from the 1960s. GPS was created and realised by the U.S. Department of Defence (DoD) and was originally run with 24 satellites.

It became fully operational in 1994. Advances in technology and new demands on the existing system have now led to efforts to modernize the GPS system and implement the next generation of GPS III satellites and Next Generation Operational Control System (OCX). Announcements from the Vice President and the White House in 1998 initiated these changes. In 2000, U.S. Congress authorised the modernisation effort, referred to as GPS III. In addition to GPS, other systems are in use or under development. The design is based partly on similar ground-based radio-navigation systems, such as LORAN and the Decca Navigator developed in the early 1940s, and used during World War II. In 1956, Friedwardt Winterberg proposed a test of general relativity (for time slowing in a strong gravitational field) using accurate atomic clocks placed in orbit inside artificial satellites. (To achieve accuracy requirements, GPS uses principles of general relativity to correct the satellites' atomic clocks.) Additional inspiration for GPS came when the Soviet Union launched the first man-made satellite, Sputnik in 1957. Two American physicists, William Guier and George Weiffenbach, at Johns Hopkins's Applied Physics Laboratory (APL), decided on their own to monitor Sputnik's radio transmissions.

First satellite navigation system, Transit (satellite), used by the United States Navy, was first successfully tested in 1960. It used a constellation of five satellites and could provide a navigational fix approximately once per hour. In 1967, the U.S. Navy developed the Timation satellite that proved the ability to place accurate clocks in space, a technology required by GPS. In the 1970s, the ground-based Omega Navigation System, based on phase comparison of signal transmission from pairs of stations, became the first worldwide radio navigation system. Limitations of these systems drove the need for a more universal navigation solution with greater accuracy. While there were wide needs for accurate navigation in military and civilian sectors, almost none of those were seen as justification for the billions of dollars it would cost in research, development, deployment, and operation for a constellation of navigation satellites.

During the Cold War arms race, the nuclear threat to the existence of the United States was the one need that did justify this cost in the view of the United States Congress. This deterrent effect is why GPS was funded. It is also the reason for the ultra secrecy at that time. The nuclear triad consisted of the United States Navy's submarine-launched ballistic missiles (SLBMs) along with United States Air Force (USAF) strategic bombers and intercontinental ballistic missiles (ICBMs). Considered vital to the nuclear deterrence posture, accurate determination of the SLBM launch position was a force multiplier. Precise navigation would enable United States submarines to get an accurate fix of their positions prior to launching their SLBMs. The USAF with two-thirds of the nuclear triad also had requirements for a more accurate and reliable navigation system.

The Navy and Air Force were developing their own technologies in parallel to solve what was essentially the same problem. To increase the survivability of ICBMs, there was a proposal to use mobile launch platforms (such as Russian SS-24 and SS-25) and so the need to fix the launch position had similarity to the SLBM situation. In 1960, the Air Force proposed a radio-navigation system called MOSAIC (MObile System for Accurate ICBM Control) that was essentially a 3-D LORAN. A follow-on study called Project 57 was worked in 1963 and it was "in this study that the GPS concept was born."

That same year the concept was pursued as Project 621B, which had "many of the attributes that you now see in GPS" and promised increased accuracy for Air Force bombers as well as ICBMs. Updates from the Navy Transit system were too slow for the high speeds of Air Force operation. The Navy Research Laboratory continued advancements with their Timation (Time Navigation) satellites, first launched in 1967, and with the third one in 1974 carrying the first atomic clock into orbit. With these parallel developments in the 1960s, it was realised that a superior system could be developed by synthesizing the best technologies from 621B, Transit, Timation, and SECOR in a multi-service program. During Labour Day weekend in 1973, a meeting of about 12 military officers at the Pentagon discussed the creation of a *Defence Navigation Satellite System (DNSS)*. It was at this meeting that "the real synthesis that became GPS was created." Later that year, the DNSS program was named *Navstar*. With the individual satellites being associated with the name Navstar (as with the predecessors Transit and Timation), a more fully encompassing name was used to identify the constellation of Navstar satellites, *Navstar-GPS*, which was later shortened simply to GPS. After Korean Air Lines Flight 007, carrying 269 people, was shot down in 1983 after straying into the USSR's prohibited airspace, in the vicinity of Sakhalin and Moneron Islands, President Ronald Reagan issued a directive making GPS freely available for civilian use, once it was sufficiently developed, as a common good. The first satellite was launched in 1989, and the 24th satellite was launched in 1994.

Initially, the highest quality signal was reserved for military use, and the signal available for civilian use was intentionally degraded (Selective Availability). This changed with President Bill Clinton ordering Selective Availability to be turned off at midnight May 1, 2000, improving the precision of civilian GPS from 100 meters (about 300 feet) to 20 meters (about 65 feet).

The executive order signed in 1996 to turn off Selective Availability in 2000 was proposed by the US Secretary of Defence, William Perry, because of the widespread growth of differential GPS services to improve civilian accuracy and eliminate the US military advantage. Moreover, the US military was actively developing technologies to deny GPS service to potential adversaries on a regional basis. Over the last decade, the U.S. has implemented several improvements to the GPS service, including new signals for civil use and increased accuracy and integrity for all users, all while maintaining compatibility with existing GPS equipment. GPS modernisation has now become an ongoing initiative to upgrade the Global Positioning System with new capabilities to meet growing military, civil, and commercial needs. The program is being implemented through a series of satellite acquisitions, including GPS Block III and the Next Generation Operational Control System (OCX). The U.S. Government continues to improve the GPS space and ground segments to increase performance and accuracy. GPS is owned and operated by the United States Government as a national resource. Department of Defence (DoD) is the steward of GPS. *Interagency GPS Executive Board (IGEB)* oversaw GPS policy matters from 1996 to 2004. After that the *National Space-Based Positioning, Navigation and Timing Executive Committee* was established by presidential directive in 2004 to advise and coordinate federal departments and agencies on matters concerning the GPS and related systems.

The executive committee is chaired jointly by the deputy secretaries of defence and transportation. Its membership includes equivalent-level officials from the departments of state, commerce, and homeland security, the joint chiefs of staff, and NASA. Components of the executive office of the president participate as observers to the executive committee, and the FCC chairman participates as a liaison. The DoD is required by law to "maintain a Standard Positioning Service (as defined in the federal radio navigation plan and the standard positioning service signal specification) that will be available on a continuous, worldwide basis," and "develop measures to prevent hostile use of GPS and its augmentations without unduly disrupting or degrading civilian uses."

Concepts of Global Positioning System

The receiver uses the messages it receives to determine the transit time of each message and computes the distance to each satellite. These distances along with the satellites' locations are used with the possible aid of trilateration, depending on which algorithm is used, to compute the position of the receiver. This position is then displayed, perhaps with a moving map display or latitude and longitude; elevation information may be included. Many GPS units show derived information such as direction and speed, calculated from position changes. Three satellites might seem enough to solve for position since space has three dimensions and a position near the Earth's surface can be assumed.

However, that would only be true if the receiver would know time extremely precisely. Even a very small clock error multiplied by the very large speed of light — the speed at which satellite signals propagate results in a large positional error. Therefore receivers use four or more satellites to solve for both the receiver's location and time. The very accurately computed time is effectively hidden by most GPS applications, which use only the location. A few specialised GPS applications do however use the time; these include time transfer, traffic signal timing, and synchronization of cell phone base stations. Although four satellites are required for normal operation, fewer apply in special cases. If one variable is already known, a receiver can determine its position using only three satellites. For example, a ship or aircraft may have known elevation.

Calculation: To provide an introductory description of how a GPS receiver works, error effects are deferred to a later section. Using messages received from a minimum of four visible satellites, a GPS receiver is able to determine the times sent and then the satellite positions corresponding to these times sent. The x, y, and z components of position, and the time sent, are designated as $[x_i, y_i, z_i, t_i]$ where the subscript i is the satellite number and has the value 1, 2, 3, or 4. Knowing the indicated time the message was received $\tilde{t}_r$, the GPS receiver could compute the transit time of the message as $(\tilde{t}_r - t_i)$, if would be equal to correct reception time, t_r. A pseudorange, $p_i \triangleq (\tilde{t}_r - t_i)c$ would be the travelling distance of the message, assuming it travelled at the speed of light, c. A satellite's position and pseudorange define a sphere, centred on the satellite, with radius equal to the pseudorange. The position of the receiver is somewhere on the surface of this sphere. Thus with four satellites, the indicated position of the GPS receiver is at or near the intersection of the surfaces of four spheres. In the ideal case of no errors, the GPS receiver would be at a precise intersection of the four surfaces. If the surfaces of two spheres intersect at more than one point, they intersect in a circle. *Two Sphere Surfaces Intersecting in a Circle*, is shown below. The distance between these two points is the diameter of the circle of intersection. The intersection of a third spherical surface with the first two will be its intersection with that circle; in most cases of practical interest, this means they intersect at two points. The two intersections are marked with dots. For automobiles and other near-earth vehicles, the correct position of the GPS receiver is the intersection closest to the Earth's surface. For space vehicles, the intersection farthest from Earth may be the correct one. The correct position for the GPS receiver is also the intersection closest to the surface of the sphere corresponding to the fourth satellite.

Space Segment

The space segment (SS) is composed of the orbiting GPS satellites, or Space Vehicles (SV) in GPS parlance. The GPS design originally called for 24 SVs, eight each in three approximately circular orbits, but this was modified to six orbital planes with four satellites each. The orbits are centred on the Earth, not rotating with the Earth, but instead fixed with respect to the distant stars. The six orbit planes have approximately 55° inclination (tilt relative to Earth's equator) and are separated by 60° right ascension of the ascending node (angle along the equator from a reference point to the orbit's intersection). The orbital period is one-half a sidereal day, i.e. 11 hours and 58 minutes. The orbits are arranged so that at least six satellites are always within line of sight from almost everywhere on Earth's surface. The result of this objective is that the four satellites are not evenly spaced (90 degrees) apart within each orbit. In general terms, the angular difference between satellites in each orbit is 30, 105, 120, and 105 degrees apart which, of course, sum to 360 degrees.

Orbiting at an altitude of approximately 20,200 km (12,600 mi); orbital radius of approximately 26,600 km (16,500 mi), each SV makes two complete orbits each sidereal day, repeating the same ground track each day. This was very helpful during development because even with only four satellites, correct alignment means all four are visible from one spot for a few hours each day. For military operations, the ground track repeat can be used to ensure good coverage in combat zones.

As of March 2008, there are 31 actively broadcasting satellites in the GPS constellation, and two older, retired from active service satellites kept in the constellation as orbital spares. The additional satellites improve the precision of GPS receiver calculations by providing redundant measurements. With the increased number of satellites, the constellation was changed to a nonuniform arrangement. Such an arrangement was shown to improve reliability and availability of the system, relative to a uniform system, when multiple satellites fail. About nine satellites are visible from any point on the ground at any one time, ensuring considerable redundancy over the minimum four satellites needed for a position.

Control Segment

The control segment is composed of:

1. a master control station (MCS),
2. an alternate master control station,
3. four dedicated ground antennas and
4. six dedicated monitor stations.

The MCS can also access U.S. Air Force Satellite Control Network (AFSCN) ground antennas (for additional command and control capability) and NGA (National Geospatial-Intelligence Agency) monitor stations. The flight paths of the satellites are tracked by dedicated U.S. Air Force monitoring stations in Hawaii, Kwajalein, Ascension Island, Diego Garcia, Colorado Springs, Colorado and Cape Canaveral, along with shared NGA monitor stations operated in England, Argentina, Ecuador, Bahrain, Australia and Washington DC. The tracking information is sent to the Air Force Space Command's MCS at Schriever Air Force Base 25 km (16 mi) ESE of Colorado Springs, which is operated by the 2nd Space Operations Squadron (2 SOPS) of the U.S. Air Force. Then 2 SOPS contacts each GPS satellite regularly with a navigational update using dedicated or shared (AFSCN) ground antennas (GPS dedicated ground antennas are located at Kwajalein, Ascension Island, Diego Garcia, and Cape Canaveral). These updates synchronize the atomic clocks on board the satellites to within a few nanoseconds of each other, and adjust the ephemeris of each satellite's internal orbital model. The updates are created by a Kalman filter that uses inputs from the ground monitoring stations, space weather information, and various other inputs.

Satellite maneuvers are not precise by GPS standards. So to change the orbit of a satellite, the satellite must be marked *unhealthy*, so receivers will not use it in their calculation. Then the maneuver can be carried out, and the resulting orbit tracked from the ground. Then the new ephemeris is uploaded and the satellite marked healthy again. The Operation Control Segment (OCS) currently serves as the control segment of record. It provides the operational capability that supports global GPS users and keeps the GPS system operational and performing within specification.

OCS successfully replaced the legacy 1970's era mainframe computer at Schriever Air Force Base in September 2007. After installation, the system helped enable upgrades and provide a foundation for a new security architecture that supported the U.S. armed forces. OCS will continue to be the ground control system of record until the new segment, Next Generation GPS Operation Control System (OCX), is fully developed and functional.

The new capabilities provided by OCX will be the cornerstone for revolutionising GPS's mission capabilities, and enabling Air Force Space Command to greatly enhance GPS operational services to U.S. combat forces, civil partners and myriad of domestic and international users. The GPS OCX program also will reduce cost, schedule and technical risk. It is designed to provide 50% sustainment cost savings through efficient software architecture and Performance-Based Logistics. In addition, GPS OCX expected to cost millions less than the cost to upgrade OCS while providing four times the capability.

The GPS OCX program represents a critical part of GPS modernisation and provides significant information assurance improvements over the current GPS OCS program.

- OCX will have the ability to control and manage GPS legacy satellites as well as the next generation of GPS III satellites, while enabling the full array of military signals.
- Built on a flexible architecture that can rapidly adapt to the changing needs of today's and future GPS users allowing immediate access to GPS data and constellations status through secure, accurate and reliable information.
- Empowers the warfighter with more secure, actionable and predictive information to enhance situational awareness.
- Enables new modernised signals (L1C, L2C, and L5) and has M-code capability, which the legacy system is unable to do.
- Provides significant information assurance improvements over the current program including detecting and preventing cyber attacks, while isolating, containing and operating during such attacks.

- Supports higher volume near real-time command and control capability.

On September 14, 2011, the U.S. Air Force announced the completion of GPS OCX Preliminary Design Review and confirmed that the OCX program is ready for the next phase of development.

The GPS OCX program has achieved major milestones and is on track to support the GPS IIIA launch in May 2014.

User Segment

The user segment is composed of hundreds of thousands of U.S. and allied military users of the secure GPS Precise Positioning Service, and tens of millions of civil, commercial and scientific users of the Standard Positioning Service. In general, GPS receivers are composed of an antenna, tuned to the frequencies transmitted by the satellites, receiver-processors, and a highly stable clock (often a crystal oscillator). They may also include a display for providing location and speed information to the user. A receiver is often described by its number of channels: this signifies how many satellites it can monitor simultaneously. Originally limited to four or five, this has progressively increased over the years so that, as of 2007, receivers typically have between 12 and 20 channels. GPS receivers may include an input for differential corrections, using the RTCM SC-104 format. This is typically in the form of an RS-232 port at 4,800 bit/s speed. Data is actually sent at a much lower rate, which limits the accuracy of the signal sent using RTCM. Receivers with internal DGPS receivers can outperform those using external RTCM data. As of 2006, even low-cost units commonly include Wide Area Augmentation System (WAAS) receivers.

GPS Receivers

Many GPS receivers can relay position data to a PC or other device using the NMEA 0183 protocol. Although this protocol is officially defined by the National Marine Electronics Association (NMEA), references to this protocol have been compiled from public records, allowing open source tools like gpsd to read the protocol without violating intellectual property laws. Other proprietary protocols exist as well, such as the SiRF and MTK protocols. Receivers can interface with other devices using methods including a serial connection, USB, or Bluetooth.

Applications

While originally a military project, GPS is considered a *dual-use* technology, meaning it has significant military and civilian applications.

GPS has become a widely deployed and useful tool for commerce, scientific uses, tracking, and surveillance. GPS's accurate time facilitates everyday activities such as banking, mobile phone operations, and even the control of power grids by allowing well synchronized hand-off switching.

Message Format

Each GPS satellite continuously broadcasts a *navigation message* at a rate of 50 bits per second. Each complete message is composed of 30 second frames, distinct groupings of 1,500 bits of information. Each frame is further subdivided into 5 subframes of length 6 seconds and with 300 bits each. Each subframe contains 10 words of 30 bits with length 0.6 seconds each. Each 30 second frame begins precisely on the minute or half minute as indicated by the atomic clock on each satellite. The first part of the message encodes the week number and the time within the week, as well as the data about the health of the satellite. The second part of the message, the *ephemeris*, provides the precise orbit for the satellite. The last part of the message, the *almanac*, contains coarse orbit and status information for all satellites in the network as well as data related to error correction.

All satellites broadcast at the same frequencies. Signals are encoded using code division multiple access (CDMA) allowing messages from individual satellites to be distinguished from each other based on unique encodings for each satellite (that the receiver must be aware of). Two distinct types of CDMA encodings are used: the coarse/ acquisition (C/A) code, which is accessible by the general public, and the precise (P) code, that is encrypted so that only the U.S. military can access it.

The ephemeris is updated every 2 hours and is generally valid for 4 hours, with provisions for updates every 6 hours or longer in non-nominal conditions. The almanac is updated typically every 24 hours. Additionally data for a few weeks following is uploaded in case of transmission updates that delay data upload.

Civilians Use

It is allowed for civilian use, with no restriction. There are two kind of GPS signals,

C/A code — which is the civilian signals and given acuracy about 8-15m

P code — which is military signal and only US military can use that, which is more accurate.

There are high-end civilian receivers available, called dual-frequency receivers, which uses part of P-code, not the full signal, and gives higher accuracy than single frequency receivers.

GPS is free for use. There is no subscription or monthly fees to pay. You need to buy a receiver that is capable of receiving the signals. But if you want to use add-on services like, differential correction for better accuracy, there may be additional fee.

Satellite Frequencies

All satellites broadcast at the same two frequencies, 1.57542 GHz (L1 signal) and 1.2276 GHz (L2 signal). The satellite network uses a CDMA spread-spectrum technique where the low-bitrate message data is encoded with a high-rate pseudo-random (PRN) sequence that is different for each satellite. The receiver must be aware of the PRN codes for each satellite to reconstruct the actual message data. The C/A code, for civilian use, transmits data at 1.023 million chips per second, whereas the P code, for U.S. military use, transmits at 10.23 million chips per second. The L1 carrier is modulated by both the C/A and P codes, while the L2 carrier is only modulated by the P code. The P code can be encrypted as a so-called P(Y) code that is only available to military equipment with a proper decryption key. Both the C/A and P(Y) codes impart the precise time-of-day to the user.

The L3 signal at a frequency of 1.38105 GHz is used by the United States Nuclear Detonation (NUDET) Detection System (USNDS) to detect, locate, and report nuclear detonations (NUDETs) in the Earth's atmosphere and near space. One usage is the enforcement of nuclear test ban treaties.

The L4 band at 1.379913 GHz is being studied for additional ionospheric correction. The L5 frequency band at 1.17645 GHZ was added in the process of GPS modernisation. This frequency falls into an internationally protected range for aeronautical navigation, promising little or no interference under all circumstances. The first Block IIF satellite that would provide this signal is set to be launched in 2009. The L5 consists of two carrier components that are in phase quadrature with each other. Each carrier component is bi-phase shift key (BPSK) modulated by a separate bit train. "L5, the third civil GPS signal, will eventually support safety-of-life applications for aviation and provide improved availability and accuracy."

A waiver has recently been granted to LightSquared to operate a terrestrial broadband service near the L1 band. Although LightSquared had applied for a license to operate in the 1525 to 1559 band as early as 2003 and it was put out for public comment, the FCC asked LightSquared to form a study group with the GPS community to test GPS receivers and identify issue that might arise due to the larger signal power from the LightSquared terrestrial network. The GPS community had not objected to the LightSquared (formerly MSV and SkyTerra) applications until sometime in late 2010. Testing in the first half of 2011 has demonstrated that the impact of the lower 10 MHz of spectrum is minimal to GPS devices (less than 1% of the total GPS devices are affected). The upper 10 MHz intended for use by LightSquared may have some impact on GPS devices. There is some concern that this will seriously degrade the GPS signal for many consumer uses. Aviation Week magazine reports that the latest testing (June 2011) confirms "significant jamming" of GPS by LightSquared's system.

Demodulation and Decoding

Because all of the satellite signals are modulated onto the same L1 carrier frequency, the signals must be separated after demodulation. This is done by assigning each satellite a unique binary sequence known as a Gold code. The signals are decoded after demodulation using addition of the Gold codes corresponding to the satellites monitored by the receiver.

If the almanac information has previously been acquired, the receiver picks the satellites to listen for by their PRNs, unique numbers in the range 1 through 32. If the almanac information is not in memory, the receiver enters a search mode until a lock is obtained on one of the satellites. To obtain a lock, it is necessary that there be an unobstructed line of sight from the receiver to the satellite. The receiver can then acquire the almanac and determine the satellites it should listen for. As it detects each satellite's signal, it identifies it by its distinct C/A code pattern. There can be a delay of up to 30 seconds before the first estimate of position because of the need to read the ephemeris data.

Processing of the navigation message enables the determination of the time of transmission and the satellite position at this time.

Bancroft's Method

Bancroft's method involves an algebraic as opposed to numerical method and can be used for the case of four satelites or for the case of more than four satellites. If there are four satellites then Bancroft's method provides the unique solution for the four unknowns. If there are more than four satellites then Bancroft's method provides the solution which minimizes the sum of the squares of the errors for the over determined system.

Trilateration

The receiver can use trilateration and one dimensional numerical root finding. Trilateration is used to determine the position based on three satellite's pseudoranges. In the usual case of two intersections, the point nearest the surface of the sphere corresponding to the fourth satellite is chosen. Let d denote the signed distance from the receiver position to the sphere around the fourth satellite. The notation, *d(correction)* shows this as a function of the correction, because it changes the pseudoranges. The problem is to determine the correction such that *d(correction)* = 0. This is the familiar problem of finding the zeroes of a one dimensional non-linear function of a scalar variable. Iterative numerical methods, such as those found in the chapter on root finding in *Numerical Recipes* can solve this type of problem.

Multidimensional Newton-Raphson Calculations

Alternatively, multidimensional root finding method such as Newton-Raphson method can be used. The approach is to linearize around an approximate solution, say $\left[x^{(k)}, y^{(k)}, z^{(k)}, b^{(k)}\right]$ from iteration k, then solve the linear equations derived from the quadratic equations above to obtain $\left[x^{(k+1)}, y^{(k+1)}, z^{(k+1)}, b^{(k+1)}\right]$ Although there is no guarantee that the method always converges due to the fact that multidimensional roots cannot be bounded, when a neighbourhood containing a solution is known as is usually the case for GPS, it is quite likely that a solution will be found. It has been shown that results are comparable in accuracy to those of the Bancroft's method.

Modernization of GPS

In 1972, the USAF Central Inertial Guidance Test Facility (Holloman AFB), conducted developmental flight tests of two prototype GPS receivers over White Sands Missile Range, using ground-based pseudo-satellites.

In 1978, the first experimental Block-I GPS satellite was launched.

In 1983, after Soviet interceptor aircraft shot down the civilian airliner KAL 007 that strayed into prohibited airspace because of navigational errors, killing all 269 people on board, U.S. President Ronald Reagan announced that GPS would be made available for civilian uses once it was completed., although it had been previously published [in Navigation magazine] that the CA code would be available to civilian users.

By 1985, ten more experimental Block-I satellites had been launched to validate the concept. Command & Control of these satellites had moved from Onizuka AFS, CA and turned over to the 2nd Satellite Control Squadron (2SCS) located at Falcon Air Force Station in Colorado Springs, Colorado.

On February 14, 1989, the first modern Block-II satellite was launched.

The Gulf War from 1990 to 1991, was the first conflict where GPS was widely used.

In 1992, the 2nd Space Wing, which originally managed the system, was de-activated and replaced by the 50th Space Wing.

By December 1993, GPS achieved initial operational capability (IOC), indicating a full constellation (24 satellites) was available and providing the Standard Positioning Service (SPS).

Full Operational Capability (FOC) was declared by Air Force Space Command (AFSPC) in April 1995, signifying full availability of the military's secure Precise Positioning Service (PPS).

In 1996, recognising the importance of GPS to civilian users as well as military users, U.S. President Bill Clinton issued a policy directive declaring GPS to be a dual-use system and establishing an Interagency GPS Executive Board to manage it as a national asset.

In 1998, United States Vice President Al Gore announced plans to upgrade GPS with two new civilian signals for enhanced user accuracy and reliability, particularly with respect to aviation safety and in 2000 the United States Congress authorised the effort, referring to it as *GPS III*.

On May 2, 2000 "Selective Availability" was discontinued as a result of the 1996 executive order, allowing users to receive a non-degraded signal globally.

In 2004, the United States Government signed an agreement with the European Community establishing cooperation related to GPS and Europe's planned Galileo system.

In 2004, United States President George W. Bush updated the national policy and replaced the executive board with the National Executive Committee for Space-Based Positioning, Navigation, and Timing.

November 2004, QUALCOMM announced successful tests of assisted GPS for mobile phones.

In 2005, the first modernised GPS satellite was launched and began transmitting a second civilian signal (L2C) for enhanced user performance.

On September 14, 2007, the aging mainframe-based Ground Segment Control System was transferred to the new Architecture Evolution Plan.

On May 19, 2009, the United States Government Accountability Office issued a report warning that some GPS satellites could fail as soon as 2010.

On May 21, 2009, the Air Force Space Command allayed fears of GPS failure saying "There's only a small risk we will not continue to exceed our performance standard."

On January 11, 2010, an update of ground control systems caused a software incompatibility with 8000 to 10000 military receivers manufactured by a division of Trimble Navigation Limited of Sunnyvale, Calif.

On February 25, 2010, the U.S. Air Force awarded the contract to develop the GPS Next Generation Operational Control System (OCX) to improve accuracy and availability of GPS navigation signals, and serve as a critical part of GPS modernisation.

A GPS satellite was launched on May 28, 2010. The oldest GPS satellite still in operation was launched on November 26, 1990, and became operational on December 10, 1990.

The GPS satellite, GPS IIF-2, was launched on July 16, 2011 at 06:41 GMT from Space Launch Complex 37B at the Cape Canaveral Air Force Station.

Standard Positioning Service (SPS)

Civil users worldwide use the SPS without charge or restrictions. Most receivers are capable of receiving and using the SPS signal. The SPS accuracy is intentionally degraded by the DOD by the use of Selective Availability.

- SPS Predictable Accuracy
- 100 meter horizontal accuracy
- 156 meter vertical accuracy
- 340 nanoseconds time accuracy

These GPS accuracy figures are from the 1999 Federal Radionavigation Plan. The figures are 95% accuracies, and express the value of two standard deviations of radial error from the actual antenna position to an ensemble of position estimates made under specified satellite elevation angle (five degrees) and PDOP (less than six) conditions.

For horizontal accuracy figures 95% is the equivalent of 2drms (two-distance root-mean-squared), or twice the radial error standard deviation. For vertical and time errors 95% is the value of two-standard deviations of vertical error or time error.

Receiver manufacturers may use other accuracy measures. Root-mean-square (RMS) error is the value of one standard deviation (68%) of the error in one, two or three dimensions. Circular Error Probable (CEP) is the value of the radius of a circle, centered at the actual position that contains 50% of the position estimates. Spherical Error Probable (SEP) is the spherical equivalent of CEP, that is the radius of a sphere, centered at the actual position, that contains 50% of the three dimension position estimates. As opposed to 2drms, drms, or RMS figures, CEP and SEP are not affected by large blunder errors making them an overly optimistic accuracy measure

Some receiver specification sheets list horizontal accuracy in RMS or CEP and without Selective Availability, making those receivers appear more accurate than those specified by more responsible vendors using more conservative error measures.

GPS Data

The GPS Navigation Message consists of time-tagged data bits marking the time of transmission of each subframe at the time they are transmitted by the SV. A data bit frame consists of 1500 bits divided into five 300-bit subframes. A data frame is transmitted every thirty seconds. Three six-second subframes contain orbital and clock data. SV Clock corrections are sent in subframe one and precise SV orbital data sets (ephemeris data parameters) for the transmitting SV are sent in subframes two and three. Subframes four and five are used to transmit different pages of system data. An entire set of twenty-five frames (125 subframes) makes up the complete Navigation Message that is sent over a 12.5 minute period.

Data frames (1500 bits) are sent every thirty seconds. Each frame consists of five subframes.

Data bit subframes (300 bits transmitted over six seconds) contain parity bits that allow for data checking and limited error correction.

Navigation Data Bits

Clock data parameters describe the SV clock and its relationship to GPS time.

Ephemeris data parameters describe SV orbits for short sections of the satellite orbits. Normally, a receiver gathers new ephemeris data each hour, but can use old data for up to four hours without much error. The ephemeris parameters are used with an algorithm that computes the SV position for any time within the period of the orbit described by the ephemeris parameter set.

- Sample Ephemeris and Clock Data Parameters
- SV Ephemeris Parameter to SV Position Algorithm
- SV Clock Parameter to SV Clock Correction Algorithm

Almanacs are approximate orbital data parameters for all SVs. The ten-parameter almanacs describe SV orbits over extended periods of time (useful for months in some cases) and a set for all SVs is sent by each SV over a period of 12.5 minutes (at least). Signal acquisition time on receiver start-up can be significantly aided by the availability of current almanacs. The approximate orbital data is used to preset the receiver with the approximate position and carrier Doppler frequency (the frequency shift caused by the rate of change in range to the moving SV) of each SV in the constellation.

Sample Almanac Parameters

Each complete SV data set includes an ionospheric model that is used in the receiver to approximates the phase delay through the ionosphere at any location and time.

Sample Ionospheric Parameters

Each SV sends the amount to which GPS Time is offset from Universal Coordinated Time. This correction can be used by the receiver to set UTC to within 100 ns.

Sample UTC Parameters

Other system parameters and flags are sent that characterize details of the system.

Position, and Time from GPS

Code Phase Tracking (Navigation)

The GPS receiver produces replicas of the C/A and/or P (Y)-Code. Each PRN code is a noise-like, but pre-determined, unique series of bits. The receiver produces the C/A code sequence for a specific SV with some form of a C/A code generator. Modern receivers usually store a complete set of precomputed C/A code chips in memory, but a hardware, shift register, implementation can also be used.

C/A Code Generator

The C/A code generator produces a different 1023 chip sequence for each phase tap setting. In a shift register implementation the code chips are shifted in time by slewing the clock that controls the shift registers. In a memory lookup scheme the required code chips are retrieved from memory.

C/A Code Phase Assignments

The C/A code generator repeats the same 1023-chip PRN-code sequence every millisecond. PRN codes are defined for 32 satellite identification numbers.

C/A Code PRN Chips

The receiver slides a replica of the code in time until there is correlation with the SV code.

Correlation Animation (250k)

Short PRN Code Segment: If the receiver applies a different PRN code to an SV signal there is no correlation.

No PRN Correlation: When the receiver uses the same code as the SV and the codes begin to line up, some signal power is detected.

Partial PRN Correlation: As the SV and receiver codes line up completely, the spread-spectrum carrier signal is de-spread and full signal power is detected.

Full PRN Correlation: A GPS receiver uses the detected signal power in the correlated signal to align the C/A code in the receiver with the code in the SV signal. Usually a late version of the code is compared with an early version to insure that the correlation peak is tracked.

Simplified GPS Receiver Block Diagram: A phase locked loop that can lock to either a positive or negative half-cycle (a bi-phase lock loop) is used to demodulate the 50 HZ navigation message from the GPS carrier signal. The same loop can be used to measure and track the carrier frequency (Doppler shift) and by keeping track of the changes to the numerically controlled oscillator, carrier frequency phase can be tracked and measured.

Data Bit Demodulation and C/A Code Control: The receiver PRN code start position at the time of full correlation is the time of arrival (TOA) of the SV PRN at receiver. This TOA is a measure of the range to SV offset by the amount to which the receiver clock is offset from GPS time. This TOA is called the pseudo-range.

Pseudo-Range Navigation

The position of the receiver is where the pseudo-ranges from a set of SVs intersect.

Intersection of Range Spheres

Position is determined from multiple pseudo-range measurements at a single measurement epoch. The pseudo range measurements are used together with SV position estimates based on the precise orbital elements (the ephemeris data) sent by each SV. This orbital data allows the receiver to compute the SV positions in three dimensions at the instant that they sent their respective signals.

Four satellites (normal navigation) can be used to determine three position dimensions and time. Position dimensions are computed by the receiver in Earth-Centered, Earth-Fixed X, Y, Z (ECEF XYZ) coordinates.

ECEF X, Y, and Z

Time is used to correct the offset in the receiver clock, allowing the use of an inexpensive receiver clock.

SV Position in XYZ is computed from four SV pseudo-ranges and the clock correction and ephemeris data.

GPS SV and Receiver XYZ

Receiver position is computed from the SV positions, the measured pseudo-ranges (corrected for SV clock offsets, ionospheric delays, and relativistic effects), and a receiver position estimate (usually the last computed receiver position).

Pseudo-Range Navigation Solution Example

Ephemeris Data Set Used in Pseudo-Range Navigation Solution Example: Three satellites could be used determine three position dimensions with a perfect receiver clock. In practice this is rarely possible and three SVs are used to compute a two-dimensional, horizontal fix (in latitude and longitude) given an assumed height. This is often possible at sea or in altimeter equipped aircraft.

Five or more satellites can provide position, time and redundancy. More SVs can provide extra position fix certainty and can allow detection of out-of-tolerance signals under certain circumstances.

Receiver Position, Velocity, and Time

Position in XYZ is converted within the receiver to geodetic latitude, longitude and height above the ellipsoid.

- Geodetic Coordinates
- ECEF XYZ to Geodetic Coordinate Conversion
- Geodetic to ECEF XYZ Coordinate Conversion

Latitude and longitude are usually provided in the geodetic datum on which GPS is based (WGS-84). Receivers can often be set to convert to other user-required datums. Position offsets of hundreds of meters can result from using the wrong datum.

Velocity is computed from change in position over time, the SV Doppler frequencies, or both.

Time is computed in SV Time, GPS Time, and UTC.

SV Time is the time maintained by each satellite. Each SV contains four atomic clocks (two cesium and two rubidium). SV clocks are monitored by ground control stations and occasionally reset to maintain time to within one-millisecond of GPS time. Clock correction data bits reflect the offset of each SV from GPS time.

SV Time is set in the receiver from the GPS signals. Data bit subframes occur every six seconds and contain bits that resolve the Time of Week to within six seconds. The 50 Hz data bit stream is aligned with the C/A code transitions so that the arrival time of a data bit edge (on a 20 millisecond interval) resolves the pseudo-range to the nearest millisecond. Approximate range to the SV resolves the twenty millisecond ambiguity, and the C/A code measurement represents time to fractional milliseconds. Multiple SVs and a navigation solution (or a known position for a timing receiver) permit SV Time to be set to an accuracy limited by the position error and the pseudo-range error for each SV.

SV Time is converted to GPS Time in the receiver.

SV Time to GPS Time Data Bits

GPS Time is a "paper clock" ensemble of the Master Control Clock and the SV clocks. GPS Time is measured in weeks and seconds from 24:00:00, January 5, 1980 and is steered to within one microsecond of UTC. GPS Time has no leap seconds and is ahead of UTC by several seconds.

GPS Week Number Rollover Comments

Time in Universal Coordinated Time (UTC) is computed from GPS Time using the UTC correction parameters sent as part of the navigation data bits.

At the transition between 23:59:59 UTC on December 31, 1998 and 00:00:00 UTC on January 1, 1999, UTC was retarded by one-second. GPS Time is now ahead of UTC by 13 seconds.

Carrier Phase Tracking (Surveying)

Carrier-phase tracking of GPS signals has resulted in a revolution in land surveying. A line of sight along the ground is no longer necessary for precise positioning. Positions can be measured up to 30 km from reference point without intermediate points. This use of GPS requires specially equipped carrier tracking receivers.

The L1 and/or L2 carrier signals are used in carrier phase surveying. L1 carrier cycles have a wavelength of 19 centimeters. If tracked and measured these carrier signals can provide ranging measurements with relative accuracies of millimeters under special circumstances. Tracking carrier phase signals provides no time of transmission information. The carrier signals, while modulated with time tagged binary codes, carry no time-tags that distinguish one cycle from another. The measurements used in carrier phase tracking are differences in carrier phase cycles and fractions of cycles over time. At least two receivers track carrier signals at the same time. Ionospheric delay differences at the two receivers must be small enough to insure that carrier phase cycles are properly accounted for. This usually requires that the two receivers be within about 30 km of each other.

Carrier phase is tracked at both receivers and the changes in tracked phase are recorded over time in both receivers.

Carrier Phase Tracking

All carrier-phase tracking is differential, requiring both a reference and remote receiver tracking carrier phases at the same time. Unless the reference and remote receivers use L1-L2 differences to measure the ionospheric delay, they must be close enough to insure that the ionospheric delay difference is less than a carrier wavelength. Using L1-L2 ionospheric measurements and long measurement averaging periods, relative positions of fixed sites can be determined over baselines of hundreds of kilometers. Phase difference changes in the two receivers are reduced using software to differences in three position dimensions between the reference station and the remote receiver. High accuracy range difference measurements with sub-centimeter accuracy are possible. Problems result from the difficulty of tracking carrier signals in noise or while the receiver moves.

Two receivers and one SV over time result in single differences.

Single Difference Survey

Two receivers and two SVs over time provide double differences.

Post processed static carrier-phase surveying can provide 1-5 cm relative positioning within 30 km of the reference receiver with measurement time of 15 minutes for short baselines (10 km) and one hour for long baselines (30 km). Rapid static or fast static surveying can provide 4-10 cm accuracies with 1 kilometer baselines and 15 minutes of recording time.

Real-Time-Kinematic (RTK) surveying techniques can provide centimeter measurements in real time over 10 km baselines tracking five or more satellites and real-time radio links between the reference and remote receivers.

GPS Error Sources

GPS errors are a combination of noise, bias, blunders.

Noise, Bias, and Blunders

Noise errors are the combined effect of PRN code noise (around 1 meter) and noise within the receiver noise (around 1 meter). Bias errors result from Selective Availability and other factors.

Selective Availability (SA)

SA is the intentional degradation of the SPS signals by a time varying bias. SA is controlled by the DOD to limit accuracy for non-U. S. military and government users. The potential accuracy of the C/A code of around 30 meters is reduced to 100 meters (two standard deviations).

The SA bias on each satellite signal is different, and so the resulting position solution is a function of the combined SA bias from each SV used in the navigation solution. Because SA is a changing bias with low frequency terms in excess of a few hours, position solutions or individual SV pseudo-ranges cannot be effectively averaged over periods shorter than a few hours. Differential corrections must be updated at a rate less than the correlation time of SA (and other bias errors).

Other Bias Error sources;

SV clock errors uncorrected by Control Segment can result in one meter errors.

Ephemeris data errors: 1 meter

Tropospheric delays: 1 meter. The troposphere is the lower part (ground level to from 8 to 13 km) of the atmosphere that experiences the changes in temperature, pressure, and humidity associated with weather changes. Complex models of tropospheric delay require estimates or measurements of these parameters.

Unmodeled ionosphere delays: 10 meters. The ionosphere is the layer of the atmosphere from 50 to 500 km that consists of ionized air.

The transmitted model can only remove about half of the possible 70 ns of delay leaving a ten meter un-modeled residual.

Multipath: 0.5 meters. Multipath is caused by reflected signals from surfaces near the receiver that can either interfere with or be mistaken for the signal that follows the straight line path from the satellite. Multipath is difficult to detect and sometime hard to avoid.

Blunders can result in errors of hundred of kilometers.

Control segment mistakes due to computer or human error can cause errors from one meter to hundreds of kilometers.

User mistakes, including incorrect geodetic datum selection, can cause errors from 1 to hundreds of meters.

Receiver errors from software or hardware failures can cause blunder errors of any size.

Noise and bias errors combine, resulting in typical ranging errors of around fifteen meters for each satellite used in the position solution.

Differential GPS (DGPS) Techniques

The idea behind all differential positioning is to correct bias errors at one location with measured bias errors at a known position. A reference receiver, or base station, computes corrections for each satellite signal.

Because individual pseudo-ranges must be corrected prior to the formation of a navigation solution, DGPS implementations require software in the reference receiver that can track all SVs in view and form individual pseudo-range corrections for each SV. These corrections are passed to the remote, or rover, receiver which must be capable of applying these individual pseudo-range corrections to each SV used in the navigation solution. Applying a simple position correction from the reference receiver to the remote receiver has limited effect at useful ranges because both receivers would have to be using the same set of SVs in their navigation solutions and have identical GDOP terms (not possible at different locations) to be identically affected by bias errors.

Differential Code GPS (Navigation)

Differential corrections may be used in real-time or later, with post-processing techniques.

Real-time corrections can be transmitted by radio link. The U. S. Coast Guard maintains a network of differential monitors and transmits DGPS corrections over radiobeacons covering much of the U. S. coastline. DGPS corrections are often transmitted in a standard format specified by the Radio Technical Commission Marine (RTCM).

Corrections can be recorded for post processing. Many public and private agencies record DGPS corrections for distribution by electronic means.

Private DGPS services use leased FM sub-carrier broadcasts, satellite links, or private radio-beacons for real-time applications. To remove Selective Availability (and other bias errors), differential corrections should be computed at the reference station and applied at the remote receiver at an update rate that is less than the correlation time of SA. Suggested DGPS update rates are usually less than twenty seconds. DGPS removes common-mode errors, those errors common to both the reference and remote receivers (not multipath or receiver noise). Errors are more often common when receivers are close together (less than 100 km). Differential position accuracies of 1-10 meters are possible with DGPS based on C/A code SPS signals.

GPS Techniques and Project Costs

Receiver costs vary depending on capabilities. Small civil SPS receivers can be purchased for under $200, some can accept differential corrections. Receivers that can store files for post-procesing with base station files cost more ($2000-5000). Receivers that can act as DGPS reference receivers (computing and providing correction data) and carrier phase tracking receivers (and two are often required) can cost many thousands of dollars ($5,000 to $40,000). Military PPS receivers may cost more or be difficult to obtain.

Other costs include the cost of multiple receivers when needed, post-processing software, and the cost of specially trained personnel.

Project tasks can often be categorized by required accuracies which will determine equipment cost.

- Low-cost, single-receiver SPS projects (100 meter accuracy)
- Medium-cost, differential SPS code Positioning (1-10 meter accuracy)
- High-cost, single-receiver PPS projects (20 meter accuracy)
- High-cost, differential carrier phase surveys (1 mm to 1 cm accuracy)

GPS Tracking Service

GPS Integrated provides Real Time Vehicle Tracking & Fleet Management Services. The complete software has been designed and developed in-house and services are supported by our trained R&D and Support teams.

GPS Integrated is based on Satellite imagery from Google Earth (http://earth.google.com) and maps from Google maps (http:// maps.google.com). You can track your vehicle in near real time (max. 5 mins delay) on your desktop using Google Earth and in web browser using Google maps.

GPS Tracking device acquire GPS signals from GPS satellites and calculates its position on the earth. This position data is sent to the gpsintegrated.com server using GPRS connection made by a integrated GSM-GPRS modem using idea cellular or airtel SIM card. This process is repeated every minute. In case GPRS connectivity is not available temporarily , the device will record the position in its memory and send it later when connectivity is established. Also in case the GPRS connectivity is not available for longer duration, it can send the position data by SMS to the server. In this way the position data of the vehicle is continuously recorded by our server gpsintegrated.com.

Every customer is provided with web based login account to access the data on the gpsintegrated.com server. You can track one or more vehicles using single account. On login our dashboard page displays important information about the vehicle like its status, max speed, motion hours, Distance covered, stationery hours. It provide you a Live Tracking link for Google Earth or Google Maps. In case you have installed free software from google called Google Earth on your computer, then just click on the Google Earth Live Tracking Link in your account and you can see the complete track of your vehicle for today along with fantastic Google earth satellite imagery.

Google Earth Satellite Imagery for most cities (like top 20 cities) in India is available with very good resolution and you can see roads and buildings clearly and identify them. New feature offered by google called Google community inside Google Earth allow you to see placemarks and names of various buildings, roads and key places in the city. In case you are on move or do not have access to your computer with google earth, you can still see the same imagery and additional maps using our browser based Google maps system. visit http:// info.gpsintegrated.com/support/maps/ to see some examples of the Google Earth and Google maps.

Along with Google earth and Google maps, we have provided various reports for vehicle owners like Start-stop report which gives information about when the vehice started, where it stopped, for how long and approx distance covered on a particular day. Day Begin-End report give more details about the vehicle from early moorning to late night. Datailed activity report gives detailed information about movements of the vehicle in a day. Other reports like Speed report, Fleet Summary report are also available. We also give customers access to detailed transaction history of the vehicle which can be exported to pdf or excel files for their further analysis.

Customers can also use Google Earth to create list of placemarks in the city like office, store, suppliers and customer locations where the vehicle normally travels. If these locations are marked in Google earth and placemarks provided to our system, then in reports you can see, when did your vehicle reached these locations and how much time they were parked at these locations.

Overall we have developed a system that is useful for individuals, corporates and travel transport companies as well as Fleet owners. Individuals can use it to track usage of their car by other family members, driver and friends.

They can also ensure security of the vehicles and recover and track them in case its stolen. We offer special manned services to track your stolen vehicles. Corporates can use our system to monitor and manage company owned vehicles, avoid misuse, increase their availability at right places at right time.

Travel transport companies and Truck fleet owners can manage their fleet of vehicles to give optimum output, avoid misuse, take quick actions in case of emmergency and more. Our System can be used in any country where GPS cellular services are available.

Chapter 5

Earth Observation Satellite

Earth observation satellites are satellites specifically designed to observe Earth from orbit, similar to reconnaissance satellites but intended for non-military uses such as environmental monitoring, meteorology, map making etc. Geostationary satellites hover over the same spot, providing continuous monitoring to a portion of the Earth's surface. Polar orbiting satellites provide global coverage, but only twice per day at any given spot.

Weather Satellite

The weather satellite is a type of satellite that is primarily used to monitor the weather and climate of the Earth. Satellites can be either polar orbiting, seeing the same swath of the Earth every 12 hours, or geostationary, hovering over the same spot on Earth by orbiting over the equator while moving at the speed of the Earth's rotation. These meteorological satellites, however, see more than clouds and cloud systems. City lights, fires, effects of pollution, auroras, sand and dust storms, snow cover, ice mapping, boundaries of ocean currents, energy flows, etc., and other types of environmental information are collected using weather satellites. Weather satellite images helped in monitoring the volcanic ash cloud from Mount St. Helens and activity from other volcanoes such as Mount Etna. Smoke from fires in the western United States such as Colorado and Utah have also been monitored.

Other environmental satellites can detect changes in the Earth's vegetation, sea state, ocean colour, and ice fields. For example, the 2002 oil spill off the northwest coast of Spain was watched carefully by the European ENVISAT, which, though not a weather satellite, flies an instrument (ASAR) which can see changes in the sea surface.

El Nino and its effects on weather are monitored daily from satellite images. The Antarctic ozone hole is mapped from weather satellite data. Collectively, weather satellites flown by the U.S., Europe, India, China, Russia, and Japan provide nearly continuous observations for a global weather watch.

Observation

Observation is typically made via different 'channels' of the Electromagnetic spectrum, in particular, the Visible and Infrared portions.

Some of these channels include.

- *Visible and Near Infrared:* 0.6 mm - 1.6 mm - For recording cloud cover during the day
- *Infrared:* 3.9 mm - 7.3 mm (Water Vapour), 8.7 mm, - 13.4 mm (Thermal imaging).

History

The first weather satellite, Vanguard 2, was launched on February 17, 1959. It was designed to measure cloud cover and resistance, but a poor axis of rotation kept it from collecting a notable amount of useful data. The first weather satellite to be considered a success was TIROS-1, launched by NASA on 1 April 1960. TIROS operated for 78 days and proved to be much more successful than Vanguard 2. TIROS paved the way for the Nimbus program, whose technology and findings are the heritage of most of the Earth-observing satellites NASA and NOAA have launched since then.

Visible Spectrum

Visible-light images from weather satellites during local daylight hours are easy to interpret even by the average person; clouds, cloud systems such as fronts and tropical storms, lakes, forests, mountains, snow ice, fires, and pollution such as smoke, smog, dust and haze are readily apparent. Even wind can be determined by cloud patterns, alignments and movement from successive photos.

Infrared Spectrum

The thermal or infrared images recorded by sensors called scanning radiometers enable a trained analyst to determine cloud heights and types, to calculate land and surface water temperatures, and to locate ocean surface features. Infrared satellite imagery can be used effectively for tropical cyclones with a visible eye pattern, using the Dvorak technique, where the difference between the temperature of the warm eye and the surrounding cold cloud tops can be used to determine its intensity (colder cloud tops generally indicate a more intense storm).

Infrared pictures depict ocean eddies or vortices and map currents such as the Gulf Stream which are valuable to the shipping industry. Fishermen and farmers are interested in knowing land and water temperatures to protect their crops against frost or increase their catch from the sea. Even El Nino phenomena can be spotted. Using colour-digitized techniques, the gray shaded thermal images can be converted to colour for easier identification of desired information.

Types

There are two basic types of meteorological satellites: geostationary and polar orbiting.

Geostationary

Geostationary weather satellites orbit the Earth above the equator at altitudes of 35,880 km (22,300 miles). Because of this orbit, they remain stationary with respect to the rotating Earth and thus can record or transmit images of the entire hemisphere below continuously with their visible-light and infrared sensors. The news media use the geostationary photos in their daily weather presentation as single images or made into movie loops. These are also available on the city forecast pages of noaa.gov (example Dallas, TX).

Several geostationary meteorological spacecraft are in operation. The United States has two in operation; GOES-11 and GOES-12. GOES-12, designated GOES-East, is located over the Amazon River and provides most of the U.S. weather information. GOES-11 is GOES-West over the eastern Pacific Ocean.

The Japanese have one in operation; MTSAT-1R over the mid Pacific at 140°E. The Europeans have Meteosat-8 (3.5°W) and Meteosat-9 (0°) over the Atlantic Ocean and have Meteosat-6 (63°E) and Meteosat-7 (57.5°E) over the Indian Ocean.

The Russians operate the GOMS over the equator south of Moscow. India also operates geostationary satellites called INSAT which carry instruments for meteorological purposes. China operated the geostationary satellites FY-2D at 86.5°E and FY-2E at 123.5°E, which are no longer in use anymore.

Polar Orbiting

Polar orbiting weather satellites circle the Earth at a typical altitude of 850 km (530 miles) in a north to south (or vice versa) path, passing over the poles in their continuous flight. Polar satellites are in sun-synchronous orbits, which means they are able to observe any place on Earth and will view every location twice each day with the same general lighting conditions due to the near-constant local solar time. Polar orbiting weather satellites offer a much better resolution than their geostationary counterparts due their closeness to the Earth.

The United States has the NOAA series of polar orbiting meteorological satellites, presently NOAA 17 and NOAA 18 as primary spacecraft, NOAA 15 and NOAA 16 as secondary spacecraft, NOAA 14 in standby, and NOAA 12. Europe has the Metop-A satellite. Russia has the Meteor and RESURS series of satellites. China has FY-1D and FY-3A. India has polar orbiting satellites as well.

DMSP

The United States Department of Defence's Meteorological Satellite (DMSP) can "see" the best of all weather vehicles with its ability to detect objects almost as 'small' as a huge oil tanker. In addition, of all the weather satellites in orbit, only DMSP can "see" at night in the visual. Some of the most spectacular photos have been recorded by the night visual sensor; city lights, volcanoes, fires, lightning, meteors, oil field burn-offs, as well as the Aurora Borealis and Aurora Australis have been captured by this 450-mile-high space vehicle's low moonlight sensor.

At the same time, energy monitoring as well as city growth can be accomplished since both major and even minor cities, as well as highway lights, are conspicuous. This informs astronomers of light pollution. The New York City Blackout of 1977 was captured by one of the night orbiter DMSP space vehicles.

In addition to monitoring city lights, these photos are a life saving asset in the detection and monitoring of fires. Not only do the satellites see the fires visually day and night, but the thermal and infrared scanners on board these weather satellites detect potential fire sources below the surface of the Earth where smoldering occurs. Once the fire is detected, the same weather satellites provide vital information about wind that could fan or spread the fires. These same cloud photos from space tell the firefighter when it will rain.

Dramatic photos are provided by all the weather satellites, but even more definitive were the DMSP night visible-light pictures of the 700 oil well fires that Iraq started on 23 February 1991 as they fled Kuwait. These fires were vividly illustrated as huge flashes in the night photos, far outstripping the glow of large populated areas. The fires consumed millions of gallons of oil; the last was doused on November 6.

Uses

Snowfield monitoring, especially in the Sierra Nevada, can be helpful to the hydrologist keeping track of how much snow is available for runoff vital to the water sheds of the western United States. This information is gleaned from existing satellites of all agencies of the U.S. government (in addition to local, on-the-ground measurements). Ice floes, packs and bergs can also be located and tracked from weather space craft.

Even pollution whether it's nature-made or man-made can be pinpointed. The visual and infrared photos show effects of pollution from their respective areas over the entire earth.

Aircraft and rocket pollution, as well as condensation trails, can also be spotted. The ocean current and low level wind information gleaned from the space photos can help predict oceanic oil spill coverage and movement. Almost every summer, sand and dust from the Sahara Desert in Africa drifts across the equatorial regions of the Atlantic Ocean.

GOES-EAST photos enable meteorologists to observe, track and forecast this sand cloud. In addition to reducing visibilities and causing respiratory problems, sand clouds suppress hurricane formation by modifying the solar radiation balance of the tropics. Other dust storms in Asia and mainland China are common and easy to spot and monitor, with recent examples of dust moving across the Pacific ocean and reaching North America.

In remote areas of the world with few local observers, fires could rage out of control for days or even weeks and consume millions of acres before authorities are alerted. Weather satellites can be a tremendous asset in such situations. Nighttime photos also clearly show the burn-off in the gas and oil fields of the Middle East and African countries. This burn-off throws large amounts of carbon dioxide into the atmosphere.

Environmental Monitoring

Other environmental satellites can assist environmental monitoring by detecting changes in the Earth's vegetation, sea state, ocean colour, and ice fields. By monitoring vegetation changes over time, droughts can be monitored by comparing the current vegetation state to its long term average. For example, the 2002 oil spill off the northwest coast of Spain was watched carefully by the European ENVISAT, which, though not a weather satellite, flies an instrument (ASAR) which can see changes in the sea surface.

Mapping

Terrain can be mapped from space with the use of satellites, such as RADARSAT-1 and TerraSAR-X.

Asdic

In 1916, under the British Board of Invention and Research, Canadian physicist Robert William Boyle took on the active sound detection project with A B Wood, producing a prototype for testing in mid 1917. This work, for the Anti-Submarine Division of the British Naval Staff, was undertaken in utmost secrecy, and used quartz piezoelectric crystals to produce the world's first practical underwater active sound detection apparatus. To maintain secrecy no mention of sound experimentation or quartz was made- the word used to describe the early work ('supersonics') was changed to 'ASD'ics, and the quartz material to 'ASD'ivite: hence the British acronym *ASDIC*. In 1939, in response to a question from the Oxford English Dictionary, the Admiralty made up the story that it stood for 'Allied Submarine Detection Investigation Committee', and this is still widely believed, though no committee bearing this name has been found in the Admiralty archives.

By 1918, both France and Britain had built prototype active systems. The British tested their ASDIC on HMS *Antrim* in 1920, and started production in 1922. The 6th Destroyer Flotilla had ASDIC-equipped vessels in 1923. An anti-submarine school, HMS *Osprey*, and a training flotilla of four vessels were established on Portland in 1924. The US Sonar QB set arrived in 1931.

By the outbreak of World War II, the Royal Navy had five sets for different surface ship classes, and others for submarines, incorporated into a complete anti-submarine attack system. The effectiveness of early ASDIC was hamstrung by the use of the depth charge as an anti-submarine weapon.

This required an attacking vessel to pass over a submerged contact before dropping charges over the stern, resulting in a loss of ASDIC contact in the moments leading up to attack. The hunter was effectively firing blind, during which time a submarine commander could take evasive action.

This situation was remedied by using several ships cooperating and by the adoption of "ahead throwing weapons", such as Hedgehog and later Squid, which projected warheads at a target ahead of the attacker and thus still in ASDIC contact. Developments during the war resulted in British ASDIC sets which used several different shapes of beam, continuously covering blind spots. Later, acoustic torpedoes were used.

At the start of World War II, British ASDIC technology was transferred for free to the United States. Research on ASDIC and underwater sound was expanded in the UK and in the US. Many new types of military sound detection were developed. These included sonobuoys, first developed by the British in 1944 under the codename *High Tea*, dipping/dunking sonar and mine detection sonar. This work formed the basis for post war developments related to countering the nuclear submarine. Work on sonar had also been carried out in the Axis countries, notably in Germany, which included countermeasures. At the end of World War II this German work was assimilated by Britain and the US. Sonars have continued to be developed by many countries, including Russia, for both military and civil uses. In recent years the major military development has been the increasing interest in low frequency active systems.

Sonar

Sonar (originally an acronym for Sound Navigation And Ranging) is a technique that uses sound propagation (usually underwater, as in Submarine navigation) to navigate, communicate with or detect other vessels. Two types of technology share the name "sonar": *passive* sonar is essentially listening for the sound made by vessels; *active* sonar is emitting pulses of sounds and listening for echoes. Sonar may be used as a means of acoustic location and of measurement of the echo characteristics of "targets" in the water.

Acoustic location in air was used before the introduction of radar. Sonar may also be used in air for robot navigation, and SODAR (an upward looking in-air sonar) is used for atmospheric investigations. The term *sonar* is also used for the equipment used to generate and receive the sound. The acoustic frequencies used in sonar systems vary from very low (infrasonic) to extremely high (ultrasonic). The study of underwater sound is known as underwater acoustics or hydroacoustics.

History

Although some animals (dolphins and bats) have used sound for communication and object detection for millions of years, use by humans in the water is initially recorded by Leonardo Da Vinci in 1490: a tube inserted into the water was said to be used to detect vessels by placing an ear to the tube.

In the 19th century an underwater bell was used as an ancillary to lighthouses to provide warning of hazards.

The use of sound to 'echo locate' underwater in the same way as bats use sound for aerial navigation seems to have been prompted by the *Titanic* disaster of 1912. The world's first patent for an underwater echo ranging device was filed at the British Patent Office by English meteorologist Lewis Richardson a month after the sinking of the Titanic, and a German physicist Alexander Behm obtained a patent for an echo sounder in 1913.

The Canadian engineer Reginald Fessenden, while working for the Submarine Signal Company in Boston, built an experimental system beginning in 1912, a system later tested in Boston Harbor, and finally in 1914 from the U.S. Revenue (now Coast Guard) Cutter Miami on the Grand Banks off Newfoundland Canada. In that test, Fessenden demonstrated depth sounding, underwater communications (Morse Code) and echo ranging (detecting an iceberg at two miles (3 km) range). The so-called Fessenden oscillator, at ca. 500 Hz frequency, was unable to determine the bearing of the berg due to the 3 metre wavelength and the small dimension of the transducer's radiating face (less than 1 metre in diameter). The ten Montreal-built British H class submarines launched in 1915 were equipped with a Fessenden oscillator.

During World War I the need to detect submarines prompted more research into the use of sound. The British made early use of underwater hydrophones, while the French physicist Paul Langevin, working with a Russian immigrant electrical engineer, Constantin Chilowski, worked on the development of active sound devices for detecting submarines in 1915 using quartz. Although piezoelectric and magnetostrictive transducers later superseded the electrostatic transducers they used, this work influenced future designs. Lightweight sound-sensitive plastic film and fibre optics have been used for hydrophones (acousto-electric transducers for in-water use), while Terfenol-D and PMN (lead magnesium niobate) have been developed for projectors.

During the 1930s American engineers developed their own underwater sound detection technology and important discoveries were made, such as thermoclines, that would help future development. After technical information was exchanged between the two countries during the Second World War, Americans began to use the term *SONAR* for their systems, coined as the equivalent of RADAR.

Performance Factors

The detection, classification and localisation performance of a sonar depends on the environment and the receiving equipment, as well as the transmitting equipment in an active sonar or the target radiated noise in a passive sonar.

Sound Propagation

Sonar operation is affected by variations in sound speed, particularly in the vertical plane. Sound travels more slowly in fresh water than in sea water, though the difference is small. The speed is determined by the water's bulk modulus and mass density. The bulk modulus is affected by temperature, dissolved impurities (usually salinity), and pressure. The density effect is small. The speed of sound (in feet per second) is approximately:

4388 + (11.25 × temperature (in °F)) + (0.0182 × depth (in feet)) + salinity (in parts-per-thousand).

This empirically derived approximation equation is reasonably accurate for normal temperatures, concentrations of salinity and the range of most ocean depths. Ocean temperature varies with depth, but at between 30 and 100 meters there is often a marked change, called the thermocline, dividing the warmer surface water from the cold, still waters that make up the rest of the ocean. This can frustrate sonar, because a sound originating on one side of the thermocline tends to be bent, or refracted, through the thermocline. The thermocline may be present in shallower coastal waters. However, wave action will often mix the water column and eliminate the thermocline. Water pressure also affects sound propagation: higher pressure increases the sound speed, which causes the sound waves to refract away from the area of higher sound speed. The mathematical model of refraction is called Snell's law.

If the sound source is deep and the conditions are right, propagation may occur in the 'deep sound channel'. This provides extremely low propagation loss to a receiver in the channel. This is because of sound trapping in the channel with no losses at the boundaries. Similar propagation can occur in the 'surface duct' under suitable conditions. However in this case there are reflection losses at the surface.

In shallow water propagation is generally by repeated reflection at the surface and bottom, where considerable losses can occur.

Sound propagation is affected by absorption in the water itself as well as at the surface and bottom. This absorption depends upon frequency, with several different mechanisms in sea water. Long-range sonar uses low frequencies to minimise absorption effects.

The sea contains many sources of noise that interfere with the desired target echo or signature. The main noise sources are waves and shipping. The motion of the receiver through the water can also cause speed-dependent low frequency noise.

Scattering

When active sonar is used, scattering occurs from small objects in the sea as well as from the bottom and surface. This can be a major source of interference. This acoustic scattering is analogous to the scattering of the light from a car's headlights in fog: a high-intensity pencil beam will penetrate the fog to some extent, but broader-beam headlights emit much light in unwanted directions, much of which is scattered back to the observer, overwhelming that reflected from the target ("white-out"). For analogous reasons active sonar needs to transmit in a narrow beam to minimise scattering.

Target Characteristics

The sound *reflection* characteristics of the target of an active sonar, such as a submarine, are known as its target strength. A complication is that echoes are also obtained from other objects in the sea such as whales, wakes, schools of fish and rocks. Passive sonar detects the target's *radiated* noise characteristics. The radiated spectrum comprises a continuous spectrum of noise with peaks at certain frequencies which can be used for classification.

Countermeasures

Active (powered) countermeasures may be launched by a submarine under attack to raise the noise level, provide a large false target, and obscure the signature of the submarine itself.

Passive (i.e., non-powered) countermeasures include:

- Mounting noise-generating devices on isolating devices.
- Sound-absorbent coatings on the hulls of submarines, for example anechoic tiles.

Active Sonar

Active sonar uses a sound transmitter and a receiver. When the two are in the same place it is monostatic operation. When the transmitter and receiver are separated it is bistatic operation. When more transmitters (or more receivers) are used, again spatially separated, it is multistatic operation. Most sonars are used monostatically with the same array often being used for transmission and reception. Active sonobuoy fields may be operated multistatically.

Active sonar creates a pulse of sound, often called a "ping", and then listens for reflections (echo) of the pulse. This pulse of sound is generally created electronically using a sonar Projector consisting of a signal generator, power amplifier and electro-acoustic transducer/array. A beamformer is usually employed to concentrate the acoustic power into a beam, which may be swept to cover the required search angles. Generally, the electro-acoustic transducers are of the Tonpilz type and their design may be optimised to achieve maximum efficiency over the widest bandwidth, in order to optimise performance of the overall system. Occasionally, the acoustic pulse may be created by other means, e.g. (1) chemically using explosives, or (2) airguns or (3) plasma sound sources.

To measure the distance to an object, the time from transmission of a pulse to reception is measured and converted into a range by knowing the speed of sound. To measure the bearing, several hydrophones are used, and the set measures the relative arrival time to each, or with an array of hydrophones, by measuring the relative amplitude in beams formed through a process called beamforming. Use of an array reduces the spatial response so that to provide wide cover multibeam systems are used. The target signal (if present) together with noise is then passed through various forms of signal processing, which for simple sonars may be just energy measurement. It is then presented to some form of decision device that calls the output either the required signal or noise. This decision device may be an operator with headphones or a display, or in more sophisticated sonars this function may be carried out by software. Further processes may be carried out to classify the target and localise it, as well as measuring its velocity.

The pulse may be at constant frequency or a chirp of changing frequency (to allow pulse compression on reception). Simple sonars generally use the former with a filter wide enough to cover possible Doppler changes due to target movement, while more complex ones generally include the latter technique. Since digital processing became available pulse compression has usually been implemented using digital correlation techniques. Military sonars often have multiple beams to provide all-round cover while simple ones only cover a narrow arc, although the beam may be rotated, relatively slowly, by mechanical scanning.

Particularly when single frequency transmissions are used, the Doppler effect can be used to measure the radial speed of a target. The difference in frequency between the transmitted and received signal is measured and converted into a velocity. Since Doppler shifts can be introduced by either receiver or target motion, allowance has to be made for the radial speed of the searching platform.

One useful small sonar is similar in appearance to a waterproof flashlight. The head is pointed into the water, a button is pressed, and the device displays the distance to the target. Another variant is a "fishfinder" that shows a small display with shoals of fish. Some civilian sonars (which are not designed for stealth) approach active military sonars in capability, with quite exotic three-dimensional displays of the area near the boat.

When active sonar is used to measure the distance from the transducer to the bottom, it is known as echo sounding. Similar methods may be used looking upward for wave measurement.

Active sonar is also used to measure distance through water between two sonar transducers or a combination of a hydrophone (underwater acoustic microphone) and projector (underwater acoustic speaker). A transducer is a device that can transmit and receive acoustic signals ("pings"). When a hydrophone/transducer receives a specific interrogation signal it responds by transmitting a specific reply signal. To measure distance, one transducer/projector transmits an interrogation signal and measures the time between this transmission and the receipt of the other transducer/hydrophone reply. The time difference, scaled by the speed of sound through water and divided by two, is the distance between the two platforms. This technique, when used with multiple transducers/hydrophones/projectors, can calculate the relative positions of static and moving objects in water.

In combat situations, an active pulse can be detected by an opponent and will reveal a submarine's position.

A very directional, but low-efficiency, type of sonar (used by fisheries, military, and for port security) makes use of a complex nonlinear feature of water known as non-linear sonar, the virtual transducer being known as a *parametric array*.

Project Artemis

Project Artemis was a one-of-a-kind low-frequency sonar for surveillance that was deployed off Bermuda for several years in the early 1960s. The active portion was deployed from a World War II tanker, and the receiving array was a built into a fixed position on an offshore bank.

Transponder

This is an active sonar device that receives a stimulus and immediately (or with a delay) retransmits the received signal or a predetermined one.

Performance Prediction

A sonar target is small relative to the sphere, centred around the emitter, on which it is located. Therefore, the power of the reflected signal is very low, several orders of magnitude less than the original signal. Even if the reflected signal was of the same power, the following example (using hypothetical values) shows the problem: Suppose a sonar system is capable of emitting a 10,000 W/m^2 signal at 1 m, and detecting a 0.001 W/m^2 signal. At 100 m the signal will be 1 W/m^2 (due to the inverse-square law). If the entire signal is reflected from a 10 m^2 target, it will be at 0.001 W/m^2 when it reaches the emitter, i.e. just detectable. However, the original signal will remain above 0.001 W/m^2 until 300 m. Any 10 m^2 target between 100 and 300 m using a similar or better system would be able to detect the pulse but would not be detected by the emitter. The detectors must be very sensitive to pick up the echoes. Since the original signal is much more powerful, it can be detected many times further than twice the range of the sonar (as in the example).

In active sonar there are two performance limitations, due to noise and reverberation. In general one or other of these will dominate so that the two effects can be initially considered separately.

In noise limited conditions at initial detection:

$$SL - 2TL + TS - (NL - DI) = DT$$

Where SL is the source level, TL is the transmission loss (or propagation loss), TS is the target strength, NL is the noise level, DI is the directivity index of the array (an approximation to the array gain) and DT is the detection threshold. In reverberation limited conditions at initial detection (neglecting array gain):

$$SL - 2TL + TS = RL + DT$$

Where RL is the reverberation level and the other factors are as before.

Marine Mammals

Active sonar may harm marine animals, although the precise mechanisms for this are not well understood. Some marine animals, such as whales and dolphins, use echolocation systems, sometimes called *biosonar* to locate predators and prey. It is conjectured that active sonar transmitters could confuse these animals and interfere with basic biological functions such as feeding and mating.

Hand-held sonar for use by a diver:

- The LIMIS (= Limpet Mine Imaging Sonar) is a hand-held or ROV-mounted imaging sonar for use by a diver. Its name is because it was designed for patrol divers (combat frogmen or Clearance Divers) to look for limpet mines in low visibility water. Links:

- Abstract of article by the International Society for Optical Engineering
- Used to find debris from the Space Shuttle Columbia crash
- Used in fish passage research at hydropower facilities

- The LUIS (= Lensing Underwater Imaging System) is another imaging sonar for use by a diver. Links:
  - Used for counting salmon in a river
- There is or was a small flashlight-shaped handheld sonar for divers, that merely displays range.
- For the INSS = Integrated Navigation Sonar System see:
  - an image.
  - short description
  - description.

Passive Sonar

Passive sonar listens without transmitting. It is often employed in military settings, although it is also used in science applications, *e.g.*, detecting fish for presence/absence studies in various aquatic environments- see also passive acoustics and passive radar. In the very broadest usage, this term can encompass virtually any analytical technique involving remotely generated sound, though it is usually restricted to techniques applied in an aquatic environment.

Identifying Sound Sources

Passive sonar has a wide variety of techniques for identifying the source of a detected sound. For example, U.S. vessels usually operate 60 Hz alternating current power systems. If transformers or generators are mounted without proper vibration insulation from the hull or become flooded, the 60 Hz sound from the windings can be emitted from the submarine or ship. This can help to identify its nationality, as most European submarines have 50 Hz power systems. Intermittent sound sources (such as a wrench being dropped) may also be detectable to passive sonar. Until fairly recently, an experienced trained operator identified signals, but now computers may do this.

Passive sonar systems may have large sonic databases, but the sonar operator usually finally classifies the signals manually. A computer system frequently uses these databases to identify classes of ships, actions (i.e. the speed of a ship, or the type of weapon released), and even particular ships. Publications for classification of sounds are provided by and continually updated by the US Office of Naval Intelligence.

Noise Limitations

Passive sonar on vehicles is usually severely limited because of noise generated by the vehicle. For this reason, many submarines operate nuclear reactors that can be cooled without pumps, using silent convection, or fuel cells or batteries, which can also run silently. Vehicles' propellers are also designed and precisely machined to emit minimal noise. High-speed propellers often create tiny bubbles in the water, and this cavitation has a distinct sound.

The sonar hydrophones may be towed behind the ship or submarine in order to reduce the effect of noise generated by the watercraft itself. Towed units also combat the thermocline, as the unit may be towed above or below the thermocline. The display of most passive sonars used to be a two-dimensional waterfall display. The horizontal direction of the display is bearing. The vertical is frequency, or sometimes time. Another display technique is to colour-code frequency-time information for bearing. More recent displays are generated by the computers, and mimic radar-type plan position indicator displays.

Performance Prediction

Unlike active sonar, only one way propagation is involved. Because of the different signal processing used, the minimum detectable signal to noise ratio will be different. The equation for determining the performance of a passive sonar is:

$$SL - TL = NL - DI + DT$$

where SL is the source level, TL is the transmission loss, NL is the noise level, DI is the directivity index of the array (an approximation to the array gain) and DT is the detection threshold. The figure of merit of a passive sonar is:

$$FOM = SL + DI - (NL + DT).$$

Warfare

Modern naval warfare makes extensive use of both passive and active sonar from water-borne vessels, aircraft and fixed installations. The relative usefulness of active versus passive sonar depends on the radiated noise characteristics of the target, generally a submarine. Although in WW II active sonar was used by surface craft—submarines avoided emitting pings which revealed their presence and position—with the advent of modern signal-processing passive sonar became preferred for initial detection. Submarines were then designed for quieter operation, and active sonar is now more used. In 1987 a division of Japanese company Toshiba reportedly sold machinery to the Soviet

Union that allowed it to mill submarine propeller blades so that they became radically quieter, creating a huge security issue with their newer generation of submarines.

Active sonar gives the exact bearing to a target, and sometimes the range. Active sonar works the same way as radar: a signal is emitted. The sound wave then travels in many directions from the emitting object. When it hits an object, the sound wave is then reflected in many other directions. Some of the energy will travel back to the emitting source. The echo will enable the sonar system or technician to calculate, with many factors such as the frequency, the energy of the received signal, the depth, the water temperature, the position of the reflecting object, etc. Active sonar is used when the platform commander determines that it is more important to determine the position of a possible threat submarine than it is to conceal his own position. With surface ships it might be assumed that the threat is already tracking the ship with satellite data. Any vessel around the emitting sonar will detect the emission. Having heard the signal, it is easy to identify the sonar equipment used (usually with its frequency) and its position (with the sound wave's energy). Active sonar is similar to radar in that, while it allows detection of targets at a certain range, it also enables the emitter to be detected at a far greater range, which is undesirable.

Since active sonar reveals the presence and position of the operator, and does not allow exact classification of targets, it is used by fast (planes, helicopters) and by noisy platforms (most surface ships) but rarely by submarines. When active sonar is used by surface ships or submarines, it is typically activated very briefly at intermittent periods to minimise the risk of detection. Consequently active sonar is normally considered a backup to passive sonar. In aircraft, active sonar is used in the form of disposable sonobuoys that are dropped in the aircraft's patrol area or in the vicinity of possible enemy sonar contacts.

Passive sonar has several advantages. Most importantly, it is silent. If the target radiated noise level is high enough, it can have a greater range than active sonar, and allows the target to be identified. Since any motorized object makes some noise, it may in principle be detected, depending on the level of noise emitted and the ambient noise level in the area, as well as the technology used. To simplify, passive sonar "sees" around the ship using it. On a submarine, nose-mounted passive sonar detects in directions of about 270°, centred on the ship's alignment, the hull-mounted array of about 160° on each side, and the towed array of a full 360°. The invisible areas are due to the ship's own interference. Once a signal is detected in a certain direction (which means that something makes sound in that direction, this is called broadband detection) it is possible to zoom in and analyse the signal received (narrowband analysis). This is generally done using a Fourier transform to show the different frequencies making up the sound. Since every engine makes a specific sound, it is straightforward to identify the object. Databases of unique engine sounds are part of what is known as *acoustic intelligence* or ACINT.

Another use of passive sonar is to determine the target's trajectory. This process is called Target Motion Analysis (TMA), and the resultant "solution" is the target's range, course, and speed. TMA is done by marking from which direction the sound comes at different times, and comparing the motion with that of the operator's own ship. Changes in relative motion are analysed using standard geometrical techniques along with some assumptions about limiting cases.

Passive sonar is stealthy and very useful. However, it requires high-tech electronic components and is costly. It is generally deployed on expensive ships in the form of arrays to enhance detection. Surface ships use it to good effect; it is even better used by submarines, and it is also used by airplanes and helicopters, mostly to a "surprise effect", since submarines can hide under thermal layers. If a submarine's commander believes he is alone, he may bring his boat closer to the surface and be easier to detect, or go deeper and faster, and thus make more sound.

Examples of sonar applications in military use are given below. Many of the civil uses given in the following section may also be applicable to naval use.

Anti-submarine Warfare

Until recently, ship sonars were usually with hull mounted arrays, either amidships or at the bow. It was soon found after their initial use that a means of reducing flow noise was required. The first were made of canvas on a framework, then steel ones were used. Now domes are usually made of reinforced plastic or pressurised rubber. Such sonars are primarily active in operation. An example of a conventional hull mounted sonar is the SQS-56.

Because of the problems of ship noise, towed sonars are also used. These also have the advantage of being able to be placed deeper in the water. However, there are limitations on their use in shallow water. These are called towed arrays (linear) or variable depth sonars (VDS) with 2/3D arrays. A problem is that the winches required to deploy/ recover these are large and expensive. VDS sets are primarily active in operation while towed arrays are passive. An example of a modern active/passive ship towed sonar is Sonar 2087 made by Thales Underwater Systems.

Torpedoes

Modern torpedoes are generally fitted with an active/passive sonar. This may be used to home directly on the target, but wake following torpedoes are also used. An early example of an acoustic homer was the Mark 37 torpedo.

Torpedo countermeasures can be towed or free. An early example was the German Sieglinde device while the Pillenwerfer was a chemical device. A widely used US device was the towed Nixie while MOSS submarine simulator was a free device. A modern alternative to the Nixie system is the UK Royal Navy S2170 Surface Ship Torpedo Defence system.

Mines

Mines may be fitted with a sonar to detect, localize and recognize the required target. Further information is given in acoustic mine and an example is the CAPTOR mine.

Mine Countermeasures

Mine Countermeasure (MCM) Sonar, sometimes called “Mine and Obstacle Avoidance Sonar (MOAS)”, is a specialised type of sonar used for detecting small objects. Most MCM sonars are hull mounted but a few types are VDS design. An example of a hull mounted MCM sonar is the Type 2193 while the SQQ-32 Mine-hunting sonar and Type 2093 systems are VDS designs.

Submarine Navigation

Submarine navigation underwater requires special skills and technologies not needed by surface ships. The challenges of underwater navigation have become more important as submarines spend more time underwater, travelling greater distances and at higher speed.

Military submarines travel underwater in an environment of total darkness with neither windows nor lights. Operating in stealth mode, they cannot use their active sonar systems to ping ahead for underwater hazards such as undersea mountains, drilling rigs or other submarines. Surfacing to obtain navigational fixes is precluded by pervasive anti-submarine warfare detection systems such as radar and satellite surveillance.

Antenna masts and antenna-equipped periscopes can be raised to obtain navigational signals but in areas of heavy surveillance, only for a few seconds or minutes; current radar technology can detect even a slender periscope while submarine shadows may be plainly visible from the air. Surfaced submarines entering and leaving port navigate similarly to traditional ships but with a few extra considerations because most of the ship rides below the waterline, making them hard for other ships to see and identify.

Navigational Technologies

Surface and Near-surface Navigation

On the surface or at periscope depth, submarines have used these methods to fix their position:

- Satellite navigation:
  - — Global positioning system (GPS)- - by entering waypoints internally, able to navigate at a more precise level.
  - — NAVSAT
- Terrestrial radio-based navigation systems; largely superseded by satellite systems
  - — LORAN — seldom if rarely used anymore.
  - — CHAYKA, the Russian counterpart of LORAN
  - — OMEGA, the Western counterpart of the Alpha Navigation System, no longer in use
  - — Alpha, the Russian counterpart of the Omega Navigation System
- Celestial navigation using the periscope, or sextant—seldom used anymore due to advancement in technology
- Radar navigation; radar signals are easily detected so radar is normally only used in friendly waters entering and exiting ports. With the implementation a more advanced radar system, many new techniques have been implemented in this process.
- Active sonar; like radar, active sonar systems are readily detected, so active sonar is usually used only entering and exiting ports.
- Pilotage — in coastal and internal waters, surfaced submarines rely on the standard system of navigational aids (buoys, navigational markers, lighthouses, etc.), utilizing the periscopes for obtaining lines of position to plot a triangulation fix.
- Voyage Management System — referred to as the VMS, utilizes digital charts with other external sources fed in, to establish the ship's position. Other information may also be entered in manually in establishing a high quality fix or position.

Deep Water Navigation

At depths below periscope depth submarines determine their position using:

- Dead reckoning course information obtained from the ship's gyrocompass, measured speed and estimates of local ocean currents, this could also be considered an estimated position as long as the ocean current is computed in.
- Inertial navigation system — this is an estimated position source, utilizes acceleration, deceleration, and pitch and roll in computing.
- Bottom contour navigation may be used in areas where detailed hydrographic data has been charted and there is adequate variation in sea floor topography. Fathometer depth measurements are compared to charted depth patterns.

Aircraft

Helicopters can be used for antisubmarine warfare by deploying fields of active/passive sonobuoys or can operate dipping sonar, such as the AQS-13. Fixed wing aircraft can also deploy sonobuoys and have greater endurance and capacity to deploy them. Processing from the sonobuoys or dipping sonar can be on the aircraft or on ship. Helicopters have also been used for mine countermeasure missions using towed sonars such as the AQS-20A.

Underwater Communications

Dedicated sonars can be fitted to ships and submarines for underwater communication.

Ocean Surveillance

For many years, the United States operated a large set of passive sonar arrays at various points in the world's oceans, collectively called Sound Surveillance System (SOSUS) and later Integrated Undersea Surveillance System (IUSS). A similar system is believed to have been operated by the Soviet Union. As permanently mounted arrays in the deep ocean were utilised, they were in very quiet conditions so long ranges could be achieved. Signal processing was carried out using powerful computers ashore. With the ending of the Cold War a SOSUS array has been turned over to scientific use. In the United States Navy, a special badge known as the Integrated Undersea Surveillance System Badge is awarded to those who have been trained and qualified in its operation.

Underwater Security

Sonar can be used to detect frogmen and other scuba divers. This can be applicable around ships or at entrances to ports. Active sonar can also be used as a deterrent and/or disablement mechanism. One such device is the Cerberus system.

Hand-held Sonar

Limpet Mine Imaging Sonar (LIMIS) is a hand-held or ROV-mounted imaging sonar designed for patrol divers (combat frogmen or clearance divers) to look for limpet mines in low visibility water.

The LUIS is another imaging sonar for use by a diver.

Integrated Navigation Sonar System (INSS) is a small flashlight-shaped handheld sonar for divers that displays range.

Intercept Sonar

This is a sonar designed to detect and locate the transmissions from hostile active sonars. An example of this is the Type 2082 fitted on the British Vanguard class submarines.

Civilian Applications

Fisheries

Fishing is an important industry that is seeing growing demand, but world catch tonnage is falling as a result of serious resource problems. The industry faces a future of continuing worldwide consolidation until a point of sustainability can be reached. However, the consolidation of the fishing fleets are driving increased demands for sophisticated fish finding electronics such as sensors, sounders and sonars. Historically, fishermen have used many different techniques to find and harvest fish. However, acoustic technology has been one of the most important driving forces behind the development of the modern commercial fisheries.

Sound waves travel differently through fish than through water because a fish's air-filled swim bladder has a different density than seawater. This density difference allows the detection of schools of fish by using reflected sound. Acoustic technology is especially well suited for underwater applications since sound travels farther and faster underwater than in air. Today, commercial fishing vessels rely almost completely on acoustic sonar and sounders to detect fish. Fishermen also use active sonar and echo sounder technology to determine water depth, bottom contour, and bottom composition.

Companies such as Raymarine UK, Marport Canada, Wesmar, Furuno, Krupp, and Simrad make a variety of sonar and acoustic instruments for the deep sea commercial fishing industry. For example, net sensors take various underwater measurements and transmit the information back to a receiver onboard a vessel. Each sensor is equipped with one or more acoustic transducers depending on its specific function. Data is transmitted from the sensors using wireless acoustic telemetry and is received by a hull mounted hydrophone. The analog signals are decoded and converted by a digital acoustic receiver into data which is transmitted to a bridge computer for graphical display on a high resolution monitor.

Echo Sounding

An echo-sounder sends an acoustic pulse directly downwards to the seabed and records the returned echo. The sound pulse is generated by a transducer that emits an acoustic pulse and then "listens" for the return signal. The time for the signal to return is recorded and converted to a depth measurement by calculating the speed of sound in water. As the speed of sound in water is around 1,500 metres per second, the time interval, measured in milliseconds, between the pulse being transmitted and the echo being received, allows bottom depth and targets to be measured.

The value of underwater acoustics to the fishing industry has led to the development of other acoustic instruments that operate in a similar fashion to echo-sounders but, because their function is slightly different from the initial model of the echo-sounder, have been given different terms.

Net Location

The net sounder is an echo sounder with a transducer mounted on the headline of the net rather than on the bottom of the vessel. Nevertheless, to accommodate the distance from the transducer to the display unit, which is much greater than in a normal echo-sounder, several refinements have to be made. Two main types are available.

The first is the cable type in which the signals are sent along a cable. In this case there has to be the provision of a cable drum on which to haul, shoot and stow the cable during the different phases of the operation. The second type is the cable less net-sounder – such as Marport's Trawl Explorer- in which the signals are sent acoustically between the net and hull mounted receiver/hydrophone on the vessel. In this case no cable drum is required but sophisticated electronics are needed at the transducer and receiver. The display on a net sounder shows the distance of the net from the bottom (or the surface), rather than the depth of water as with the echo-sounder's hull-mounted transducer. Fixed to the headline of the net, the footrope can usually be seen which gives an indication of the net performance.

Any fish passing into the net can also be seen, allowing fine adjustments to be made to catch the most fish possible. In other fisheries, where the amount of fish in the net is important, catch sensor transducers are mounted at various positions on the cod-end of the net. As the cod-end fills up these catch sensor transducers are triggered one by one and this information is transmitted acoustically to display monitors on the bridge of the vessel.

The skipper can then decide when to haul the net. Modern versions of the net sounder, using multiple element transducers, function more like a sonar than an echo sounder and show slices of the area in front of the net and not merely the vertical view that the initial net sounders used.

The sonar is an echo-sounder with a directional capability that can show fish or other objects around the vessel.good

Ship Velocity Measurement

Sonars have been developed for measuring a ship's velocity either relative to the water or to the bottom.

ROV and UUV

Small sonars have been fitted to Remotely Operated Vehicles (ROV) and Unmanned Underwater Vehicles (UUV) to allow their operation in murky conditions. These sonars are used for looking ahead of the vehicle. The Long-Term Mine Reconnaissance System is an UUV for MCM purposes.

Vehicle Location

Sonars which act as beacons are fitted to aircraft to allow their location in the event of a crash in the sea. Short and Long Baseline sonars may be used for caring out the location, such as LBL.

Scientific Applications

Biomass Estimation

Detection of fish, and other marine and aquatic life, and estimation their individual sizes or total biomass using active sonar techniques. As the sound pulse travels through water it encounters objects that are of different density or acoustic characteristics than the surrounding medium, such as fish, that reflect sound back toward the sound source. These echoes provide information on fish size, location, abundance and Behaviour. Data is usually processed and analysed using a variety of software such as Echoview.

Wave Measurement

An upward looking echo sounder mounted on the bottom or on a platform may be used to make measurements of wave height and period. From this statistics of the surface conditions at a location can be derived.

Water Velocity Measurement

Special short range sonars have been developed to allow measurements of water velocity.

Bottom Type Assessment

Sonars have been developed that can be used to characterise the sea bottom into, for example, mud, sand, and gravel. Relatively simple sonars such as echo sounders can be promoted to seafloor classification systems via add-on modules, converting echo parameters into sediment type. Different algorithms exist, but they are all based on changes in the energy or shape of the reflected sounder pings. Advanced substrate classification analysis can be achieved using calibrated (scientific) echosounders and parametric or fuzzy-logic analysis of the acoustic data.

Bottom Topography Measurement

Side-scan sonars can be used to derive maps of the topography of an area by moving the sonar across it just above the bottom. Low frequency sonars such as GLORIA have been used for continental shelf wide surveys while high frequency sonars are used for more detailed surveys of smaller areas.

Sub-bottom Profiling

Powerful low frequency echo-sounders have been developed for providing profiles of the upper layers of the ocean bottom.

Synthetic Aperture Sonar

Various synthetic aperture sonars have been built in the laboratory and some have entered use in mine-hunting and search systems. An explanation of their operation is given in synthetic aperture sonar.

Parametric Sonar

Parametric sources use the non-linearity of water to generate the difference frequency between two high frequencies. A virtual end-fire array is formed. Such a projector has advantages of broad bandwidth, narrow beamwidth, and when fully developed and carefully measured it has no obvious sidelobes: see Parametric array. Its major disadvantage is very low efficiency of only a few percent.. P.J. Westervelt's seminal 1963 JASA paper summarizes the trends involved.

Seismometer

Seismometers are instruments that measure motions of the ground, including those of seismic waves generated by earthquakes, nuclear explosions, and other seismic sources. Records of seismic waves allow seismologists to map the interior of the Earth, and locate and measure the size of these different sources. It is often used to mean *seismometer*, though it is more applicable to the older instruments in which the measuring and recording of ground motion were combined than to modern systems, in which these functions are separated. Both types provide a continuous record of ground motion; this distinguishes them from seismoscopes, which merely indicate that motion has occurred, perhaps with some simple measure of how large it was.

Basic Principles

Inertial seismometers have:

- A weight, usually called the internal mass, that can move relative to the instrument frame, but is attached to it by a system (such as a spring) that will hold it fixed relative to the frame if there is no motion, and also damp out any motions once the motion of the frame stops.
- A means of recording the motion of the mass relative to the frame, or the force needed to keep it from moving.

Any motion of the ground moves the frame. The mass tends not to move because of its inertia, and by measuring the motion between the frame and the mass the motion of the ground can be determined, even though the mass does move.

Early seismometers used optical levers or mechanical linkages to amplify the small motions involved, recording on soot-covered paper or photographic paper. Modern instruments use electronics. In some systems, the mass is held nearly motionless relative to the frame by an electronic negative feedback loop. The motion of the mass relative to the frame is measured, and the feedback loop applies a magnetic or electrostatic force to keep the mass nearly motionless. The voltage needed to produce this force is the output of the seismometer, which is recorded digitally. In other systems the weight is allowed to move, and its motion produces a voltage in a coil attached to the mass and moving through the magnetic field of a magnet attached to the frame. This design is often used in the geophones used in seismic surveys for oil and gas. Professional seismic observatories usually have instruments measuring three axes: north-south, east-west, and the vertical. If only one axis can be measured, this is usually the vertical because it is less noisy and gives better records of some seismic waves.

The foundation of a seismic station is critical. A professional station is sometimes mounted on bedrock. The best mountings may be in deep boreholes, which avoid thermal effects, ground noise and tilting from weather and tides. Other instruments are often mounted in insulated enclosures on small buried piers of unreinforced concrete. Reinforcing rods and aggregates would distort the pier as the temperature changes. A site is always surveyed for ground noise with a temporary installation before pouring the pier and laying conduit.

Zhang Heng's Seismoscope

In AD 132, Zhang Heng of China's Han dynasty invented the first seismoscope (by the definition above), which was called *Houfeng Didong Yi* (literally, "instrument for measuring the seasonal winds and the movements of the Earth"). The description we have, from the History of the Later Han Dynasty, says that it was a large bronze vessel, about 2 meters in diameter; at eight points around the top were dragon's heads holding bronze balls. When there was an earthquake, one of the mouths would open and drop its ball into a bronze toad at the base, making a sound and supposedly showing the direction of the earthquake. On at least one occasion, probably at the time of a large earthquake in Gansu in 143 CE, the seismoscope indicated an earthquake even though one was not felt.

The available text says that inside the vessel was a central column that could move along eight tracks; this is thought to refer to a pendulum, though it is not known exactly how this was linked to a mechanism that would open only one dragon's mouth. The first ever earthquake recorded by this seismograph was supposedly *somewhere in the east*. Days later, a rider from the east reported this earthquake.

An Early Example

The principle can be shown by an early special purpose seismometer. This consisted of a large stationary pendulum, with a stylus on the bottom. As the earth starts to move, the heavy mass of the pendulum has the inertia to stay still in the non-earth frame of reference. The result is that the stylus scratches a pattern corresponding with the Earth's movement. This type of strong motion seismometer recorded upon a smoked glass (glass with carbon soot). While not sensitive enough to detect distant earthquakes, this instrument could indicate the direction of the pressure waves and thus help find the epicenter of a local earthquake — such instruments were useful in the analysis of the 1906 San Francisco earthquake. Further re-analysis was performed in the 1980s using these early recordings, enabling a more precise determination of the initial fault break location in Marin county and its subsequent progression, mostly to the south.

Early Designs

After 1880, most seismometers were descended from those developed by the team of John Milne, James Alfred Ewing and Thomas Gray, who worked in Japan from 1880 to 1895. These seismometers used damped horizontal pendulums. After World War II, these were adapted into the widely used Press-Ewing seismometer.

Later, professional suites of instruments for the world-wide standard seismographic network had one set of instruments tuned to oscillate at fifteen seconds, and the other at ninety seconds, each set measuring in three directions. Amateurs or observatories with limited means tuned their smaller, less sensitive instruments to ten seconds. The basic damped horizontal pendulum seismometer swings like the gate of a fence. A heavy weight is mounted on the point of a long (from 10 cm to several meters) triangle, hinged at its vertical edge. As the ground moves, the weight stays unmoving, swinging the "gate" on the hinge.

The advantage of a horizontal pendulum is that it achieves very low frequencies of oscillation in a compact instrument. The "gate" is slightly tilted, so the weight tends to slowly return to a central position. The pendulum is adjusted (before the damping is installed) to oscillate once per three seconds, or once per thirty seconds. The general-purpose instruments of small stations or amateurs usually oscillate once per ten seconds. A pan of oil is placed under the arm, and a small sheet of metal mounted on the underside of the arm drags in the oil to damp oscillations. The level of oil, position on the arm, and angle and size of sheet is adjusted until the damping is "critical," that is, almost having oscillation. The hinge is very low friction, often torsion wires, so the only friction is the internal friction of the wire. Small seismographs with low proof masses are placed in a vacuum to reduce disturbances from air currents.

Zollner described torsionally-suspended horizontal pendulums as early as 1869, but developed them for gravimetry rather than seismometry. Early seismometers had an arrangement of levers on jeweled bearings, to scratch smoked glass or paper. Later, mirrors reflected a light beam to a direct-recording plate or roll of photographic paper. Briefly, some designs returned to mechanical movements to save money. In mid-twentieth-century systems, the light was reflected to a pair of differential electronic photosensors called a photomultiplier. The voltage generated in the photomultiplier was used to drive galvanometers which had a small mirror mounted on the axis. The moving reflected light beam would strike the surface of the turning drum, which was covered with photo-sensitive paper. The expense of developing photo sensitive paper caused many seismic observatories to switch to ink or thermal-sensitive paper.

Another relatively simple device was used in the late 19th and early 20th century. This consisted of a pendulum free to swing in any direction, with a scribe at the bottom touching a smoked glass plate. While not providing time information or information on distant earthquakes these did give accurate initial shock directions and proved useful in a late 20th century analysis of the 1906 San Francisco earthquake.

Modern Instruments

Modern instruments use electronic sensors, amplifiers, and recording devices. Most are broadband covering a wide range of frequencies. Some seismometers can measure motions with frequencies from 500 Hz to 0.00118 Hz (1/500 = 0.002 seconds per cycle, to 1/0.00118 = 850 seconds per cycle). The mechanical suspension for horizontal instruments remains the garden-gate described above. Vertical instruments use some kind of constant-force suspension, such as the LaCoste suspension. The LaCoste suspension uses a zero-length spring to provide a long period (high sensitivity). Some modern instruments use a "triaxial" design, in which three identical motion sensors are set at the same angle to the vertical but 120 degrees apart on the horizontal. Vertical and horizontal motions can be computed from the outputs of the three sensors.

Seismometers unavoidably introduce some distortion into the signals they measure, but professionally-designed systems have carefully characterized frequency transforms. Modern sensitivities come in three broad ranges: geophones, 50 to 750 V/m; local geologic seismographs, about 1,500 V/m; and teleseismographs, used for world survey, about 20,000 V/m. Instruments come in three main varieties: short period, long period and broadband. The short and long period measure velocity and are very sensitive, however they 'clip' the signal or go off-scale for ground motion that is strong enough to be felt by people. A 24-bit analog-to-digital conversion channel is commonplace. Practical devices are linear to roughly one part per million.

Delivered seismometers come with two styles of output: analog and digital. Analog seismographs require analog recording equipment, possibly including an analog-to-digital converter. The output of a digital seismographs can be simply input to a computer. They present the data in standard digital forms (often "SE2" over ethernet).

Teleseismometers

The modern broadband seismograph can record a very broad range of frequencies. It consists of a small "proof mass", confined by electrical forces, driven by sophisticated electronics. As the earth moves, the electronics attempt to hold the mass steady through a feedback circuit. The amount of force necessary to achieve this is then recorded. In most designs the electronics holds a mass motionless relative to the frame. This device is called a "force balance accelerometer". It measures acceleration instead of velocity of ground movement. Basically, the distance between the mass and some part of the frame is measured very precisely, by a linear variable differential transformer. Some instruments use a linear variable differential capacitor.

That measurement is then amplified by electronic amplifiers attached to parts of an electronic negative feedback loop. One of the amplified currents from the negative feedback loop drives a coil very like a loudspeaker, except that the coil is attached to the mass, and the magnet is mounted on the frame. The result is that the mass stays nearly motionless.

Most instruments measure directly the ground motion using the distance sensor. The voltage generated in a sense coil on the mass by the magnet directly measures the instantaneous velocity of the ground. The current to the drive coil provides a sensitive, accurate measurement of the force between the mass and frame, thus measuring directly the ground's acceleration (using f=ma where f=force, m=mass, a=acceleration). One of the continuing problems with sensitive vertical seismographs is the buoyancy of their masses. The uneven changes in pressure caused by wind blowing on an open window can easily change the density of the air in a room enough to cause a vertical seismograph to show spurious signals. Therefore, most professional seismographs are sealed in rigid gas-tight enclosures. For example, this is why a common Streckheisen model has a thick glass base that must be glued to its pier without bubbles in the glue.

It might seem logical to make the heavy magnet serve as a mass, but that subjects the seismograph to errors when the Earth's magnetic field moves. This is also why seismograph's moving parts are constructed from a material that interacts minimally with magnetic fields. A seismograph is also sensitive to changes in temperature so many instruments are constructed from low expansion materials such as nonmagnetic invar.

The hinges on a seismograph are usually patented, and by the time the patent has expired, the design has been improved. The most successful public domain designs use thin foil hinges in a clamp. Another issue is that the transfer function of a seismograph must be accurately characterized, so that its frequency response is known. This is often the crucial difference between professional and amateur instruments. Most instruments are characterized on a variable frequency shaking table.

Strong-motion Seismometers

Another type of seismometer is a digital strong-motion seismometer, or accelerograph. The data from such an instrument is essential to understand how an earthquake affects manmade structures. A strong-motion seismometer measures acceleration. This can be mathematically integrated later to give velocity and position. Strong-motion seismometers are not as sensitive to ground motions as teleseismic instruments but they stay on scale during the strongest seismic shaking.

Other Forms

Accelerographs and geophones are often heavy cylindrical magnets with a spring-mounted coil inside. As case moves, the coil tends to stay stationary, so the magnetic field cuts the wires, inducing current in the output wires. They receive frequencies from several hundred hertz down to 4.5 Hz or even as low as 1 Hz with higher quality models. Some have electronic damping, a low-budget way to get some of the performance of the closed-loop wide-band geologic seismographs.

Strain-beam accelerometers constructed as integrated circuits are too insensitive for geologic seismographs (2002), but are widely used in geophones. Some other sensitive designs measure the current generated by the flow of a non-corrosive ionic fluid through an electret sponge or a conductive fluid through a magnetic field.

Modern Recording

Today, the most common recorder is a computer with an analog-to-digital converter, a disk drive and an internet connection; for amateurs, a PC with a sound card and associated software is adequate. Most systems record continuously, but some record only when a signal is detected, as shown by a short-term increase in the variation of the signal, compared to its long-term average (which can vary slowly because of changes in seismic noise).

Interconnected Seismometers

Seismometers spaced in an array can also be used to precisely locate, in three dimensions, the source of an earthquake, using the time it takes for seismic waves to propagate away from the hypocenter, the initiating point of fault rupture. Interconnected seismometers are also used to detect underground nuclear test explosions. These seismometer are often used as part of a large scale, multi-million dollar governmental or scientific project, but some organizations, such as the

Quake-Catcher Network, can use residential size detectors built into computers to detect earthquakes as well.

In reflection seismology, an array of seismometers image subsurface features. The data are reduced to images using algorithms similar to tomography. The data reduction methods resemble those of computer-aided tomographic medical imaging X-ray machines (CAT-scans), or imaging sonars. A world-wide array of seismometers can actually image the interior of the Earth in wave-speed and transmissivity. This type of system uses events such as earthquakes, impact events or nuclear explosions as wave sources.

The first efforts at this method used manual data reduction from paper seismograph charts. Modern digital seismograph records are better adapted to direct computer use. With inexpensive seismometer designs and internet access, amateurs and small institutions have even formed a "public seismograph network." Seismographic systems used for petroleum or other mineral exploration historically used an explosive and a wireline of geophones unrolled behind a truck. Now most short-range systems use "thumpers" that hit the ground, and some small commercial systems have such good digital signal processing that a few sledgehammer strikes provide enough signal for short-distance refractive surveys.

Exotic cross or two-dimensional arrays of geophones are sometimes used to perform three-dimensional reflective imaging of subsurface features. Basic linear refractive geomapping software (once a black art) is available off-the-shelf, running on laptop computers, using strings as small as three geophones. Some systems now come in an 18" (0.5 m) plastic field case with a computer, display and printer in the cover. Small seismic imaging systems are now sufficiently inexpensive to be used by civil engineers to survey foundation sites, locate bedrock, and find subsurface water.

To coordinate a series of large-scale observations, most sensing systems depend on the following: platform location, what time it is, and the rotation and orientation of the sensor. High-end instruments now often use positional information from satellite navigation systems. The rotation and orientation is often provided within a degree or two with electronic compasses. Compasses can measure not just azimuth (i.e. degrees to magnetic north), but also altitude (degrees above the horizon), since the magnetic field curves into the Earth at different angles at different latitudes. More exact orientations require gyroscopic-aided orientation, periodically realigned by different methods including navigation from stars or known benchmarks.

Resolution impacts collection and is best explained with the following relationship: less resolution=less detail & larger coverage, More resolution=more detail, less coverage. The skilled management of collection results in cost-effective collection and avoid situations such as the use of multiple high resolution data which tends to clog transmission and storage infrastructure.

Data Processing

Generally speaking, remote sensing works on the principle of the *inverse problem*. While the object or phenomenon of interest (the state) may not be directly measured, there exists some other variable that can be detected and measured (the observation), which may be related to the object of interest through the use of a data-derived computer model. The common analogy given to describe this is trying to determine the type of animal from its footprints. For example, while it is impossible to directly measure temperatures in the upper atmosphere, it is possible to measure the spectral emissions from a known chemical species (such as carbon dioxide) in that region.

The frequency of the emission may then be related to the temperature in that region via various thermodynamic relations. The quality of remote sensing data consists of its spatial, spectral, radiometric and temporal resolutions.

Spatial Resolution

The size of a pixel that is recorded in a raster image – typically pixels may correspond to square areas ranging in side length from 1 to 1,000 metres (3.3 to 3,300 ft).

Spectral Resolution

The wavelength width of the different frequency bands recorded – usually, this is related to the number of frequency bands recorded by the platform. Current Landsat collection is that of seven bands, including several in the infra-red spectrum, ranging from a spectral resolution of 0.07 to 2.1 μm. The Hyperion sensor on Earth Observing-1 resolves 220 bands from 0.4 to 2.5 μm, with a spectral resolution of 0.10 to 0.11 μm per band.

Radiometric Resolution

The number of different intensities of radiation the sensor is able to distinguish. Typically, this ranges from 8 to 14 bits, corresponding to 256 levels of the gray scale and up to 16,384 intensities or "shades" of colour, in each band. It also depends on the instrument noise.

Temporal Resolution

The frequency of flyovers by the satellite or plane, and is only relevant in time-series studies or those requiring an averaged or mosaic image as in deforesting monitoring. This was first used by the intelligence community where repeated coverage revealed changes in infrastructure, the deployment of units or the modification/introduction of equipment. Cloud cover over a given area or object makes it necessary to repeat the collection of said location.

In order to create sensor-based maps, most remote sensing systems expect to extrapolate sensor data in relation to a reference point including distances between known points on the ground. This depends on the type of sensor used.

For example, in conventional photographs, distances are accurate in the centre of the image, with the distortion of measurements increasing the farther you get from the centre. Another factor is that of the platen against which the film is pressed can cause severe errors when photographs are used to measure ground distances. The step in which this problem is resolved is called georeferencing, and involves computer-aided matching up of points in the image (typically 30 or more points per image) which is extrapolated with the use of an established benchmark, "warping" the image to produce accurate spatial data. As of the early 1990s, most satellite images are sold fully georeferenced.

In addition, images may need to be radiometrically and atmospherically corrected.

Radiometric Correction

Gives a scale to the pixel values, e.g. the monochromatic scale of 0 to 255 will be converted to actual radiance values.

Atmospheric Correction

Eliminates atmospheric haze by rescaling each frequency band so that its minimum value (usually realised in water bodies) corresponds to a pixel value of 0. The digitizing of data also make possible to manipulate the data by changing gray-scale values.

Interpretation is the critical process of making sense of the data. The first application was that of aerial photographic collection which used the following process; spatial measurement through the use of a light table in both conventional single or stereographic coverage, added skills such as the use of photogrammetry, the use of photomosaics, repeat coverage, Making use of objects' known dimensions in order to detect modifications. Image Analysis is the recently developed automated computer-aided application which is in increasing use.

Object-Based Image Analysis (OBIA) is a sub-discipline of GIScience devoted to partitioning remote sensing (RS) imagery into meaningful image-objects, and assessing their characteristics through spatial, spectral and temporal scale.

Old data from remote sensing is often valuable because it may provide the only long-term data for a large extent of geography. At the same time, the data is often complex to interpret, and bulky to store. Modern systems tend to store the data digitally, often with lossless compression. The difficulty with this approach is that the data is fragile, the format may be archaic, and the data may be easy to falsify. One of the best systems for archiving data series is as computer-generated machine-readable ultrafiche, usually in typefonts such as OCR-B, or as digitized half-tone images.

Ultrafiches survive well in standard libraries, with lifetimes of several centuries. They can be created, copied, filed and retrieved by automated systems. They are about as compact as archival magnetic media, and yet can be read by human beings with minimal, standardized equipment.

Chapter 6

Geomagnetic Reversal of the Earth

Geomagnetic polarity during the late Cenozoic Era. Dark areas denote periods where the polarity matches today's polarity, light areas denote periods where that polarity is reversed.

A geomagnetic reversal is a change in the orientation of Earth's magnetic field such that the positions of magnetic north and magnetic south become interchanged. These events often involve an extended decline in field strength followed by a rapid recovery after the new orientation has been established. These events occur on a scale of tens of thousands of years or longer.

History

In the early 20th century geologists first noticed that some volcanic rocks were magnetized in a direction opposite to what was expected. The first examination of the timing of magnetic reversals was done by Motonori Matuyama in the 1920s, who observed that there were rocks in Japan whose magnetic fields were reversed and those were all of early Pleistocene age or older. At the time he published his proposal suggesting that the magnetic field had been reversed, the magnetic field itself was poorly understood so there was little interest in the possibility that it had reversed.

Three decades later, theories existed of the cause of the magnetic field and some of these included the possibility of field reversal. Most paleomagnetic research in the late 1950s was examining the wandering of the poles and continental drift. Although it was discovered that some rocks would reverse their magnetic field while cooling, it became apparent that most magnetized volcanic rocks contained traces of the

Earth's magnetic field at the time that the rock cooled. At first it seemed that reversals happen every one million years, but during the 1960s it became apparent that the time between reversals is erratic.

During the 1950s and 1960s research ships gathered information about variations in the Earth's magnetic field. Because of the complex routes of cruises, associating navigational data with magnetometer readings was difficult. But when data was plotted on a map, it became apparent that there were remarkably regular and continuous magnetic stripes across the ocean floors.

In 1963 Frederick Vine and Drummond Matthews provided a simple explanation by combining the seafloor spreading theory of Harry Hess with the known time scale of reversals: if new sea floor acquired the present magnetic field, spreading from a central ridge would produce magnetic stripes parallel to the ridge. Canadian L. W. Morley independently proposed a similar explanation in January 1963, but his work was rejected by the scientific journals *Nature* and *Journal of Geophysical Research*, and not published until 1967 in the literary magazine *Saturday Review*. The Morley–Vine–Matthews hypothesis was the first key scientific test of the seafloor spreading theory of continental drift.

Starting in 1966, Lamont–Doherty Geological Observatory scientists found the magnetic profiles across the Pacific-Antarctic Ridge were symmetrical and matched the pattern in the north Atlantic's Reykjanes ridges. The same magnetic anomalies were found over most of the world's oceans, and allowed estimation of the timing of the creation of most of the oceanic crust.

Through analysis of palaeomagnetic data, it is now known that the Earth's magnetic field has reversed its orientation tens of thousands of times during the history of the Earth since its formation. With the increasingly accurate Global Polarity Timescale (GPTS) it has become apparent that the rate at which reversals occur has varied considerably throughout the past.

During some periods of geologic time (e.g. Cretaceous Long Normal), the Earth's magnetic field is observed to maintain a single orientation for tens of millions of years. Other events seem to have occurred very rapidly, with two reversals in a span of 50,000 years. Furthermore, Occidental College geologist Scott Bogue and Jonathan Glen of the US Geological Survey have found evidence in ancient lava rock in Battle Mountain, Nevada that could indicate that a rapid geomagnetic field reversal occurred over a period of four years. Another offered explanation is a rapid acceleration period within the steady movement of the field. The reversal was dated to around 15 million years ago. The last reversal was the Brunhes–Matuyama reversal approximately 780,000 years ago.

Causes

Scientific opinion is divided on what causes geomagnetic reversals. One theory holds that they are due to events internal to the system that generates the Earth's magnetic field. The other holds that they are due to external events.

Internal Events

Many scientists believe that reversals are an inherent aspect of the dynamo theory of how the geomagnetic field is generated. In computer simulations, it is observed that magnetic field lines can sometimes become tangled and disorganized through the chaotic motions of liquid metal in the Earth's core. In some simulations, this leads to an instability in which the magnetic field spontaneously flips over into the opposite orientation. This scenario is supported by observations of the solar magnetic field, which undergoes spontaneous reversals every 9–12 years. However, with the sun it is observed that the solar magnetic intensity greatly increases during a reversal, whereas all reversals on Earth seem to occur during periods of low field strength.

Present computational methods have used very strong simplifications in order to produce models that run to acceptable time scales for research programs.

External Events

Others, such as Richard A. Muller, believe that geomagnetic reversals are not spontaneous processes but rather are triggered by external events which directly disrupt the flow in the Earth's core. Such processes may include the arrival of continental slabs carried down into the mantle by the action of plate tectonics at subduction zones, the initiation of new mantle plumes from the core-mantle boundary, and possibly mantle-core shear forces resulting from very large impact events.

Supporters of this theory hold that any of these events could lead to a large scale disruption of the dynamo, effectively turning off the geomagnetic field. Because the magnetic field is stable in either the present North-South orientation or a reversed orientation, they propose that when the field recovers from such a disruption it spontaneously chooses one state or the other, such that a recovery is seen as a reversal in about half of all cases.

Brief disruptions which do not result in reversal are also known and are called geomagnetic excursions.

Observing Past Fields

Past field reversals can be and have been recorded in the "frozen" ferromagnetic (or more accurately, ferrimagnetic) minerals of solidified sedimentary deposits or cooled volcanic flows on land.

Originally, however, the past record of geomagnetic reversals was first noticed by observing the magnetic stripe "anomalies" on the ocean floor. Lawrence W. Morley, Frederick John Vine and Drummond Hoyle Matthews made the connection to seafloor spreading in the Morley-Vine-Matthews hypothesis which soon led to the development of the theory of plate tectonics. Given that the sea floor spreads at a relatively constant rate, this results in broadly evident substrate "stripes" from which the past magnetic field polarity can be inferred by looking at the data gathered from towing a magnetometer along the sea floor.

However, because no existing unsubducted sea floor (or sea floor thrust onto continental plates, such as in the case of ophiolites) is much older than about 180 million years (Ma) in age, other methods are necessary for detecting older reversals. Most sedimentary rocks incorporate tiny amounts of iron rich minerals, whose orientation is influenced by the ambient magnetic field at the time at which they formed. Under favourable conditions, it is thus possible to extract information of the variations in magnetic field from many kinds of sedimentary rocks. However, subsequent diagenetic processes after burial may erase evidence of the original field.

Because the magnetic field is present globally, finding similar patterns of magnetic variations at different sites is one method used to correlate age across different locations. In the past four decades great amounts of paleomagnetic data about seafloor ages (up to ~250 Ma) have been collected and have become an important and convenient tool to estimate the age of geologic sections. It is not an independent dating method, but is dependent on "absolute" age dating methods like radioisotopic systems to derive numeric ages. It has become especially useful to metamorphic and igneous geologists where the use of index fossils to estimate ages is seldom available.

Geomagnetic Polarity Time Scale

Changing Frequency of Geomagnetic Reversals over Time

The rate of reversals in the Earth's magnetic field has varied widely over time. 72 million years ago (Ma), the field reversed 5 times in a million years. In a 4-million-year period centered on 54 Ma, there were 10 reversals; at around 42 Ma, 17 reversals took place in the span of 3 million years. In a period of 3 million years centering on 24 Ma, 13 reversals occurred. No fewer than 51 reversals occurred in a 12-million-year period, centering on 15 million years ago. These eras of frequent reversals have been counterbalanced by a few "superchrons" – long periods when no reversals took place.

It had generally been assumed that the frequency of geomagnetic reversals is random; in 2006, a team of physicists at the University of Calabria found that the reversals conform to a Levy distribution, which describes stochastic processes with long-ranging correlations between events in time.

Cretaceous Long Normal Superchron

A long period of time during which there were no magnetic pole reversals, the Cretaceous Long Normal (also called the Cretaceous Superchron or C34) lasted for almost 40 million years, from about 120 to 83 million years ago. This time period included stages of the Cretaceous period from the Aptian through the Santonian can be seen when looking at the frequency of magnetic reversals approaching and following the Cretaceous Long Normal. The frequency steadily decreased prior to the period, reaching its low point (no reversals) during the period. Following the Cretaceous Superchron the frequency of reversals slowly increased over the next 80 million years, to the present.

Jurassic Quiet Zone

The Jurassic Quiet Zone is a section of ocean floor which is completely devoid of the magnetic stripes that can be detected elsewhere. This could mean that there was a long period of polar stability during the Jurassic period similar to the Cretaceous Superchron. Another possibility is that as this is the oldest section of ocean floor, any magnetization that did exist has completely degraded by now. The Jurassic Quiet Zones exist in places along the continental margins of the Atlantic ocean as well as in parts the Western Pacific (such as just east of the Mariana Trench).

Kiaman Long Reversed Superchron

This long period without geomagnetic reversals lasted from approximately the late Carboniferous to the late Permian, or for more than 50 million years, from around 316 to 262 million years ago. The magnetic field was reversed compared to its present state. The name "Kiaman" derives from the Australian village of Kiama, where some of the first geological evidence of the superchron was found in 1925.

Moyero Reversed Superchron

This period in the Ordovician of more than 20 million years (485 to 463 million years ago) is suspected to host another superchron (Pavlov &. Gallet 2005, Episodes, 2005). But until now this possible superchron has only been found in the Moyero river section north of the polar circle in Siberia.

Future of the Present Field

At present, the overall geomagnetic field is becoming weaker at a rate which would, if it continues, cause the dipole field to temporarily collapse by 3000–4000 CE. The South Atlantic Anomaly is believed by some scientists, including Dr. Pieter Kotze, head of the geomagnetism group at the Hermanus Magnetic Observatory in the southern Cape, to be a product of this. The present strong deterioration corresponds to a 10–15% decline over the last 150 years and has accelerated in the past several years; however, geomagnetic intensity has declined almost continuously from a maximum 35% above the modern value achieved approximately 2000 years ago. The rate of decrease and the current strength are within the normal range of variation, as shown by the record of past magnetic fields recorded in rocks.

The nature of Earth's magnetic field is one of heteroscedastic fluctuation. An instantaneous measurement of it, or several measurements of it across the span of decades or centuries, is not sufficient to extrapolate an overall trend in the field strength. It has gone up and down in the past with no apparent reason. Also, noting the local intensity of the dipole field (or its fluctuation) is insufficient to characterize Earth's magnetic field as a whole, as it is not strictly a dipole field. The dipole component of Earth's field can diminish even while the total magnetic field remains the same or increases.

The Earth's magnetic north pole is drifting from northern Canada towards Siberia with a presently accelerating rate — 10 km per year at the beginning of the 20th century, up to 40 km per year in 2003, and since then has only accelerated. Glatzmaier and collaborator Paul Roberts of UCLA have made a numerical model of the electromagnetic, fluid dynamical processes of Earth's interior, and computed it on a Cray supercomputer. The results reproduced key features of the magnetic field over more than 40,000 years of simulated time. Additionally, the computer-generated field reversed itself.

Effects on Biosphere and Human Society

Because the magnetic field has never been observed to reverse by humans with instrumentation, and the mechanism of field generation is not well understood, it is difficult to say what the characteristics of the magnetic field might be leading up to such a reversal.

Some speculate that a greatly diminished magnetic field during a reversal period will expose the surface of the Earth to a substantial and potentially damaging increase in cosmic radiation. However, *Homo erectus* and their ancestors certainly survived many previous reversals.

There is no uncontested evidence that a magnetic field reversal has ever caused any biological extinctions. A possible explanation is that the solar wind may induce a sufficient magnetic field in the Earth's ionosphere to shield the surface from energetic particles even in the absence of the Earth's normal magnetic field.

Magnetic Field Detection

The Earth's magnetic field strength was measured by Carl Friedrich Gauss in 1835 and has been repeatedly measured since then, showing a relative decay of about 10% over the last 150 years. The Magsat satellite and later satellites have used 3-axis vector magnetometers to probe the 3-D structure of the Earth's magnetic field. The later Orsted satellite allowed a comparison indicating a dynamic geodynamo in action that appears to be giving rise to an alternate pole under the Atlantic Ocean west of S. Africa.

Governments sometimes operate units that specialise in measurement of the Earth's magnetic field. These are geomagnetic observatories, typically part of a national Geological Survey, for example the British Geological Survey's Eskdalemuir Observatory. Such observatories can measure and forecast magnetic conditions that sometimes affect communications, electric power, and other human activities. The International Real-time Magnetic Observatory Network, with over 100 interlinked geomagnetic observatories around the world has been recording the earths magnetic field since 1991.

The military determines local geomagnetic field characteristics, in order to detect *anomalies* in the natural background that might be caused by a significant metallic object such as a submerged submarine. Typically, these magnetic anomaly detectors are flown in aircraft like the UK's Nimrod or towed as an instrument or an array of instruments from surface ships.

Commercially, geophysical prospecting companies also use magnetic detectors to identify naturally occurring anomalies from ore bodies, such as the Kursk Magnetic Anomaly. Animals including birds and turtles can detect the Earth's magnetic field, and use the field to navigate during migration. Cows and wild deer tend to align their bodies north-south while relaxing, but not when the animals are under high voltage power lines, leading researchers to believe magnetism is responsible. Seismo-electromagnetics is an area of research aimed at earthquake prediction.

Earth's Rotation

Earth's rotation is the rotation of the solid Earth around its own axis. The Earth rotates towards the east. As viewed from the North Star Polaris, the Earth turns counter-clockwise.

Rotation Period

On a prograde planet like the Earth, the stellar day is shorter than the solar day. At time 1, the Sun and a certain distant star are both overhead. At time 2, the planet has rotated 360° and the distant star is overhead again but the Sun is not (1→2 = one stellar day). It is not until a little later, at time 3, that the Sun is overhead again (1→3 = one solar day).

Earth's rotation period relative to the Sun (true noon to true noon) is its true solar day or apparent solar day. It is the derivative of the equation of time and thus depends on the eccentricity of Earth's orbit and the tilt (obliquity) of Earth's axis. Both vary over thousands of years so the annual variation of the true solar day also varies. Generally, it is longer than the mean solar day twice a year and shorter twice a year. The true solar day tends to be longer near perihelion when the Sun apparently moves along the ecliptic through a greater angle than usual, taking about 10 seconds longer to do so. Conversely, it is about 10 seconds shorter near aphelion. It is about 20 seconds longer near a solstice when the projection of the Sun's apparent movement along the ecliptic onto the celestial equator causes the Sun to move through a greater angle than usual. Conversely, near an equinox the projection onto the equator is shorter by about 20 seconds. Currently, the perihelion and solstice effects combine to lengthen the true solar day near December 22 by 30 mean solar seconds, but the solstice effect is partially cancelled by the aphelion effect near June 19 when it is only 13 seconds longer. The effects of the equinoxes shorten it near March 26 and September 16 by 18 seconds and 21 seconds, respectively.

The average of the true solar day over an entire year is the mean solar day, which contains 86,400 mean solar seconds. Currently, each of these seconds is slightly longer than an SI second because Earth's mean solar day is now slightly longer than it was during the 19th century due to tidal friction. The average length of the mean solar day since the introduction of the leap second in 1972 has been about 0 to 2 ms longer than 86,400 SI seconds. Random fluctuations due to core-mantle coupling have an amplitude of about 5 ms. The mean solar second between 1750 and 1892 was chosen in 1895 by Simon Newcomb as the independent unit of time in his Tables of the Sun. These tables were used to calculate the world's ephemerides between 1900 and 1983, so this second became known as the ephemeris second. The SI second was made equal to the ephemeris second in 1967.

Earth's rotation period relative to the fixed stars, called its stellar day by the International Earth Rotation and Reference Systems Service (IERS), is 86,164.098 903 691 seconds of mean solar time (UT1) (23^h 56^m 4.098 903 691^s, 0.997 269 663 237 16 mean solar days). Earth's rotation period relative to the precessing or moving mean vernal equinox, misnamed its sidereal day, is 86,164.090 530 832 88 seconds of mean solar time (UT1) (23^h 56^m 4.090 530 832 88^s, 0.997 269 566 329 08 mean solar days). Thus the sidereal day is shorter than the stellar day by about 8.4 ms. In 499 CE Aryabhata, a great mathematician-astronomer from the classical age of Indian mathematics and Indian astronomy estimated the length of sidereal day as 23^h 56^m 4.1^s.

Both the stellar day and the sidereal day are shorter than the mean solar day by about 3 minutes 56 seconds. The mean solar day in SI seconds is available from the IERS for the periods 1623–2005 and 1962–2005.

Recently (1999–2010) the average annual length of the mean solar day in excess of 86,400 SI seconds has varied between 0.25 ms and 1 ms, which must be added to both the stellar and sidereal days given in mean solar time above to obtain their lengths in SI seconds.

The angular speed of Earth's rotation in inertial space is (7.2921150 ± 0.0000001) $\times 10^{-5}$ radians per SI second (mean solar second). Multiplying by (180°/π radians)×(86,400 seconds/mean solar day) yields 360.9856°/mean solar day, indicating that Earth rotates more than 360° relative to the fixed stars in one solar day. Earth's movement along its nearly circular orbit while it is rotating once around its axis requires that Earth rotate slightly more than once relative to the fixed stars before the mean Sun can pass overhead again, even though it rotates only once (360°) relative to the mean Sun. Multiplying the value in rad/s by Earth's equatorial radius of 6,378,137 m (WGS84 ellipsoid) (factors of 2π radians needed by both cancel) yields an equatorial speed of 465.1 m/s, 1,674.4 km/h or 1,040.4 mi/h. Some sources state that Earth's equatorial speed is slightly less, or 1,669.8 km/h. This is obtained by dividing Earth's equatorial circumference by 24 hours. However, the use of only one circumference unwittingly implies only one rotation in inertial space, so the corresponding time unit must be a sidereal hour. This is confirmed by multiplying by the number of sidereal days in one mean solar day, 1.002 737 909 350 795, which yields the equatorial speed in mean solar hours given above of 1,674.4 km/h.

The permanent monitoring of the Earth's rotation requires the use of Very Long Baseline Interferometry coordinated with the Global Positioning System, Satellite laser ranging, and other satellite techniques. This provides the absolute reference for the determination of universal time, precession, and nutation.

Over millions of years, the rotation is significantly slowed by gravitational interactions with the Moon: see tidal acceleration. However some large scale events, such as the 2004 Indian Ocean earthquake, have caused the rotation to speed up by around 3 microseconds. Post-glacial rebound, ongoing since the last Ice age, is changing the distribution of the Earth's mass thus affecting the Moment of Inertia of the Earth and, by the Conservation of Angular Momentum, the Earth's rotation period.

Precession, Nutation, and Polar Motion

The Earth's rotation axis moves with respect to the fixed stars (inertial space); the components of this motion are precession and nutation. The Earth's rotation axis also moves with respect to the Earth's crust; this is called polar motion.

Precession is a rotation of the Earth's rotation axis, caused primarily by external torques from the gravity of the Sun, Moon and other bodies. The polar motion is primarily due to free core nutation and the Chandler wobble.

Physical Effects

The velocity of the rotation of Earth has had various effects over time, including the Earth's shape (an oblate spheroid), climate, ocean depth and currents, and tectonic forces.

Origin

It is theorized that Earth formed as part of the birth of the Solar System: what eventually became the solar system initially existed as a large, rotating cloud of dust, rocks, and gas. It was composed of hydrogen and helium produced in the Big Bang, as well as heavier elements ejected by supernovas. As this interstellar dust is inhomogeneous, any asymmetry during gravitational accretion results in the angular momentum of the eventual planet. The current rotation period of the

Earth is the result of this initial rotation and other factors, including tidal friction and the hypothetical impact of Theia.

Evidence of Earth's Rotation

In the Earth's rotating frame of reference, a freely moving body follows an apparent path that deviates from the one it would follow in a fixed frame of reference. Because of this Coriolis effect, falling bodies veer eastward from the vertical plumb line below their point of release, and projectiles veer right in the northern hemisphere (and left in the southern) from the direction in which they are shot.

The Coriolis effect has many other manifestations, especially in meteorology, where it is responsible for the differing rotation direction of cyclones in the northern and southern hemispheres. Hooke, following a 1679 suggestion from Newton, tried unsuccessfully to verify the predicted half millimeter eastward deviation of a body dropped from a height of 8.2 meters, but definitive results were only obtained later, in the late 18th and early 19th century, by Giovanni Battista Guglielmini in Bologna, Johann Friedrich Benzenberg in Hamburg and Ferdinand Reich in Freiberg, using taller towers and carefully released weights.

Foucault Pendulum

The most celebrated test of Earth's rotation is the Foucault pendulum first built by physicist Leon Foucault in 1851, which consisted of an iron sphere suspended 67 m from the top of the Pantheon in Paris. Because of the Earth's rotation under the swinging pendulum the pendulum's plane of oscillation appears to rotate at a rate depending on latitude. At the latitude of Paris the predicted and observed shift was about 11 degrees clockwise per hour. Foucault pendulums now swing in museums around the world.

Earth's Orbit

In astronomy, the Earth's orbit is the motion of the Earth around the Sun, at an average distance of about 150 million kilometres, every 365.256363 mean solar days (1 sidereal year). This motion gives an apparent movement of the Sun with respect to the stars at a rate of about 1°/day (or a Sun or Moon diameter every 12 hours) eastward, as seen from Earth. On average it takes 24 hours—a solar day—for Earth to complete a full rotation about its axis relative to the Sun so that the Sun returns to the meridian. The orbital speed of the Earth around the Sun averages about 30 km/s (108,000 km/h), which is fast enough to cover the planet's diameter (about 12,700 km) in seven minutes, and the distance to the Moon of 384,000 km in four hours.

Viewed from a vantage point above the north poles of both the Sun and the Earth, the Earth appears to revolve in a counterclockwise direction about the Sun. From the same vantage point both the Earth and the Sun would appear to rotate in a counterclockwise direction about their respective axes.

Heliocentrism

Heliocentrism, or heliocentricism, is the astronomical model in which the Earth and planets revolve around a stationary Sun at the Centre of the solar system. The word comes from the Greek. Historically, heliocentrism was opposed to geocentrism, which placed the Earth at the Centre. The notion that the Earth revolves around the Sun was first proposed in the 3rd century BC by Aristarchus of Samos. However, it was not until the 16th century that a fully predictive mathematical model of a heliocentric system was presented, by the Renaissance mathematician and astronomer Nicolaus Copernicus, leading to the Copernican Revolution. In the following century, this model was elaborated and expanded by Johannes Kepler and supporting observations made using a telescope were presented by Galileo Galilei. With the observations of William Herschel, astronomers realized that the sun was not the Centre of the universe and by the 1920s Edwin Hubble had shown that it was part of a galaxy that was only one of many billions.

Early Developments

To anyone who stands and looks at the sky, it seems clear that the Earth stays in one place while everything in the sky rises in the east and sets in the west once a day. Observing over a longer time, one sees more complicated movements. The Sun makes a slower circle eastward over the course of a year; the planets have similar motions, but they sometimes move in the reverse direction for a while (retrograde motion).

As these motions became better understood, more elaborate descriptions were required, the most famous of which was the geocentric Ptolemaic system, which achieved its full expression in the 2nd century. The Ptolemaic system was a sophisticated astronomical system that managed to calculate the positions for the planets to a fair degree of accuracy. Ptolemy himself, in his *Almagest*, points out that any model for describing the motions of the planets is merely a mathematical device, and since there is no actual way to know which is true, the simplest model that gets the right numbers should be used. However, he rejected the idea of a spinning earth as absurd since it would create huge winds. His planetary hypotheses were sufficiently real that the distances of moon, sun, planets and stars could be determined by treating orbits' celestial spheres as contiguous realities. This made the stars' distance less than 20 Astronomical Units, a regression, since Aristarchus of Samos's heliocentric scheme had centuries earlier necessarily placed the stars at least two orders of magnitude more distant.

Philosophical Discussions

Philosophical arguments on heliocentrism involve general statements that the Sun is at the Centre of the universe or that some or all of the planets revolve around the Sun, and arguments supporting these claims. These ideas can be found in a range of Greek, Arabic and Latin texts. These early sources, however, do not provide techniques to compute any observational consequences of their proposed heliocentric ideas.

Greek and Hellenistic World

The first non-geocentric model of the Universe was proposed by the Pythagorean philosopher Philolaus (d. 390 BC). According to Philolaus, at the Centre of the Universe was a "central fire" around which the Earth, Sun, Moon and Planets revolved in uniform circular motion. This system postulated the existence of a counter-earth collinear with the Earth and central fire, with the same period of revolution around the central fire as the Earth. The Sun revolved around the central fire once a year, and the stars were stationary. The Earth maintained the same hidden face towards the central fire, rendering both it and the "counter-earth" invisible from Earth. The Pythagorean concept of uniform circular motion remained unchallenged for approximately the next 2000 years, and it was to the Pythagoreans that Copernicus referred to show that the notion of a moving Earth was neither new nor revolutionary.

Kepler gives an alternative explanation of the Pythagoreans' "central fire" as the sun, "as most sects purposely hid[e] their teachings". Heraclides of Pontus (4th century BC) explained the apparent daily motion of the celestial sphere through the rotation of the Earth. It used to be thought that he believed Mercury and Venus to revolve around the Sun, which in turn (along with the other planets) revolves around the Earth. Ambrosius Theodosius Macrobius (395–423 AD) later described this as the "Egyptian System," stating that "it did not escape the skill of the Egyptians," though there is no other evidence it was known in ancient Egypt.

Aristarchus's 3rd century BC calculations on the relative sizes of the Earth, Sun and Moon, from a 10th century AD Greek copy.

The first person known to have proposed a heliocentric system, however, was Aristarchus of Samos (*c.* 270 BC). Like Eratosthenes, Aristarchus calculated the size of the Earth, and measured the size and distance of the Moon and Sun, in a treatise which has survived. From his estimates, he concluded that the Sun was six to seven times wider than the Earth and thus hundreds of times more voluminous. His writings on the heliocentric system are lost, but some information is known from surviving descriptions and critical commentary by his contemporaries, such as Archimedes. Some have suggested that his calculation of the relative size of the Earth and Sun led Aristarchus to conclude that it made more sense for the Earth to be moving than for the huge Sun to be moving around it. Though the original text has been lost, a reference in Archimedes' book *The Sand Reckoner* describes another work by Aristarchus in which he advanced an alternative hypothesis of the heliocentric model. Archimedes wrote:

You King Gelon are aware the 'universe' is the name given by most astronomers to the sphere the Centre of which is the Centre of the Earth, while its radius is equal to the straight line between the Centre of the Sun and the Centre of the Earth. This is the common account as you have heard from astronomers. But Aristarchus has brought out a book consisting of certain hypotheses, wherein it appears, as a consequence of the assumptions made, that the universe is many times greater than the 'universe' just mentioned. His hypotheses are that the fixed stars and the Sun remain unmoved, that the Earth revolves about the Sun on the circumference of a circle, the Sun lying in the middle of the orbit, and that the sphere of fixed stars, situated about the same Centre as the Sun, is so great that the circle in which he supposes the Earth to revolve bears such a proportion to the distance of the fixed stars as the Centre of the sphere bears to its surface.

Aristarchus thus believed the stars to be very far away, and saw this as the reason why there was no visible parallax, that is, an observed movement of the stars relative to each other as the Earth moved around the Sun. The stars are in fact much farther away than the distance that was generally assumed in ancient times, which is why stellar parallax is only detectable with telescopes.

Archimedes says that Aristarchus made the stars' distance larger, suggesting that he was answering the natural objection that heliocentrism requires stellar parallactic oscillations. He apparently agreed to the point but placed the stars so distant as to make the parallactic motion invisibly minuscule. Thus heliocentrism opened the way for realization that the universe was larger than the geocentrists taught.

Seleucus of Seleucia

It should be noted that since Plutarch mentions the 'followers of Aristarchus' in passing, there were likely other astronomers in the Classical period who also espoused heliocentrism, but whose work is now lost to us. The only other astronomer from antiquity known by name who is known to have supported Aristarchus' heliocentric model was Seleucus of Seleucia (b. 190 BC), a hellenized Babylonian astronomer who flourished a century after Aristarchus in the Seleucid empire. Seleucus adopted the heliocentric system of Aristarchus and is said to have proved the heliocentric theory. According to Bartel Leendert van der Waerden, Seleucus may have proved the heliocentric theory by determining the constants of a geometric model for the heliocentric theory and by developing methods to compute planetary positions using this model. He may have used early trigonometric methods that were available in his time, as he was a contemporary of Hipparchus. A fragment of a work by Seleucus of Seleucia, who supported Aristarchus' heliocentric model in the 2nd century BC, has survived in Arabic translation, which was referred to by Rhazes (b. 865).

Western Christendom

There were occasional speculations about heliocentrism in Europe before Copernicus. In Roman Carthage, Martianus Capella (5th century A.D.) expressed the opinion that the planets Venus and Mercury did not go about the Earth but instead circled the Sun. Capella's model was discussed in the Early Middle Ages by various anonymous 9th-century commentators and Copernicus mentions him as an influence on his own work.

During the Late Middle Ages, Bishop Nicole Oresme discussed the possibility that the Earth rotated on its axis, while Cardinal Nicholas of Cusa in his *Learned Ignorance* asked whether there was any reason to assert that the Sun (or any other point) was the Centre of the universe. In parallel to a mystical definition of God, Cusa wrote that "Thus the fabric of the world (*machina mundi*) will *quasi* have its Centre everywhere and circumference nowhere."

Mathematical Astronomy

In mathematical astronomy, models of heliocentrism involve mathematical computational systems that are tied to a heliocentric model and where positions of the planets can be derived. The first computational system explicitly tied to a heliocentric model was the Copernican model described by Copernicus, but there were earlier computational systems that may have implied some form of heliocentricity, notably Aryabhata's model, which has astronomical parameters which some have interpreted to imply a form of heliocentricity. Several Muslim astronomers also developed computational systems with astronomical parameters compatible with heliocentricity, as stated by Biruni, but the concept of heliocentrism was considered a philosophical problem rather than a mathematical problem. Their astronomical parameters may have been later adapted in the Copernican model in a heliocentric context.

Ancient India

Aryabhata (476–550), in his magnum opus *Aryabhatiya* (499), propounded a computational system based on a planetary model in which the Earth was taken to be spinning on its axis and the periods of the planets were given with respect to the Sun. B. L. van der Waerden has interpreted this to be a heliocentric model but this view has been strongly disputed by others. He accurately calculated many astronomical constants, such as the periods of the planets, times of the solar and lunar eclipses, and the instantaneous motion of the Moon. Early followers of Aryabhata's model included Varahamihira, Brahmagupta, and Bhaskara II.

Nilakantha Somayaji (1444–1544), in his *Aryabhatiyabhasya*, a commentary on Aryabhata's *Aryabhatiya*, developed a computational system for a partially heliocentric planetary model, in which the planets orbit the Sun, which in turn orbits the Earth, similar to the Tychonic system later proposed by Tycho Brahe in the late 16th century. In the *Tantrasangraha* (1500), he further revised his planetary system, which was mathematically more accurate at predicting the heliocentric orbits of the interior planets than both the Tychonic and Copernican models, but like Indian astronomy in general fell short of proposing models of the universe. Nilakantha's planetary system also incorporated the Earth's rotation on its axis. Most astronomers of the Kerala school of astronomy and mathematics seem to have accepted his planetary model.

Islamic World

Due to the scientific dominance of the Ptolemaic system in Islamic astronomy, the Muslim astronomers accepted unanimously the geocentric model. However, several Muslim scholars questioned the Earth's apparent immobility and centrality within the universe. Alhazen wrote a scathing critique of Ptolemy's model in his *Doubts on Ptolemy* (c. 1028), which some have interpreted to imply he was criticizing Ptolemy's geocentrism, but most agree that he was actually criticizing the details of Ptolemy's model rather than his geocentrism. Alhazen did, however, later propose the Earth's rotation on its axis in *The Model of the Motions* (c. 1038).

Abu Rayhan Biruni (b. 973) discussed the possibility of whether the Earth rotated about its own axis and around the Sun, but in his *Masudic Canon*, he set forth the principles that the Earth is at the Centre of the universe and that it has no motion of its own. He was aware that if the Earth rotated on its axis and around the Sun, this would be consistent with his astronomical parameters, but he considered this a philosophical problem rather than a mathematical one.

At the Maragha observatory, Najm al-Din al-Qazwini al-Katibi (d. 1277), in his *Hikmat al-'Ain*, wrote an argument for a heliocentric model, but later abandoned the model. Qotb al-Din Shirazi (b. 1236) also discussed the possibility of heliocentrism, but rejected it. Ibn al-Shatir (b. 1304) developed a geocentric system that employed mathematical techniques, such as the Tusi-couple and Urdi lemma, that were almost identical to those Nicolaus Copernicus later employed in his heliocentric system, implying that its mathematical model was influenced by the Maragha school. At the Maragha and Samarkand observatories, the Earth's rotation was discussed by Tusi (b. 1201) and Qushji (b. 1403); the arguments and evidence they used resemble those used by Copernicus to support the Earth's motion.

However, it remains a fact that the Maragha school never made the big leap to heliocentrism. In addition, the influence of the Maragha school on Copernicus remains speculative, since there is no documentary evidence to prove it. The possibility that Copernicus independently developed the Tusi couple remains open, since no researcher has yet proven that he knew about Tusi′s work or the Maragha school.

It has been argued that, given some differences between the two models, it is more likely that Copernicus could have taken the ideas found in the Tusi couple from Proclus's *Commentary on the First Book of Euclid*. Another possible source for Copernicus's knowledge is the *Questiones de Spera* of Nicole Oresme, who described how a reciprocating linear motion of a celestial body could be produced by a combination of circular motions similar to those proposed by al-Tusi.

Copernican Revolution

Astronomical Model

In the 16th century, Nicolaus Copernicus's *De revolutionibus* presented a full discussion of a heliocentric model of the universe in much the same way as Ptolemy's *Almagest* had presented his geocentric model in the 2nd century. Copernicus discussed the philosophical implications of his proposed system, elaborated it in full geometrical detail, used selected astronomical observations to derive the parameters of his model, and wrote astronomical tables which enabled one to compute the past and future positions of the stars and planets.

In doing so, Copernicus moved heliocentrism from philosophical speculation to predictive geometrical astronomy. This theory resolved the issue of planetary retrograde motion by arguing that such motion was only perceived and apparent, rather than real: it was a parallax effect, as a car that one is passing seems to move backwards against the horizon. This issue was also resolved in the geocentric Tychonic system; the latter, however, while eliminating the major epicycles, retained as a physical reality the irregular back-and-forth motion of the planets, which Kepler characterized as a "pretzel".

Copernicus cited Aristarchus in an early (unpublished) manuscript of *De Revolutionibus* (which still survives) so he was clearly aware of at least one previous proponent of the heliocentric thesis. However, in the published version he restricts himself to noting that in works by Cicero he had found an account of the theories of Hicetas and that Plutarch had provided him with an account of the Pythagoreans Heraclides Ponticus, Philolaus, and Ecphantus. These authors had proposed a moving earth, which did not, however, revolve around a central sun.

Religious Attitudes to Heliocentrism

Heliocentrism had been in conflict with religion before Copernicus. One of the few pieces of information we have about the reception of Aristarchus's heliocentric system comes from a passage in Plutarch's dialogue, *Concerning the Face which Appears in the Orb of the Moon.* According to one of Plutarch's characters in the dialogue, the philosopher Cleanthes had held that Aristarchus should be charged with impiety for "moving the hearth of the world". In fact, however, Aristarchus's heliocentrism appears to have attracted little attention, religious or otherwise, until Copernicus revived and elaborated it.

Circulation of Commentariolus (Before 1533)

The first information about the heliocentric views of Nicolaus Copernicus were circulated in manuscript. Although only in manuscript, Copernicus' ideas were well known among astronomers and others. His ideas contradicted the then-prevailing understanding of the Bible. In the King James Bible Chronicles 16:30 state that "the world also shall be stable, that it be not moved." Psalm 104:5 says, "[the Lord] Who laid the foundations of the earth, that it should not be removed for ever." Ecclesiastes 1:5 states that "The sun also ariseth, and the sun goeth down, and hasteth to his place where he arose."

Nonetheless, in 1533, Johann Albrecht Widmannstetter delivered in Rome a series of lectures outlining Copernicus' theory. The lectures were heard with interest by Pope Clement VII and several Catholic cardinals. On 1 November 1536, Archbishop of Capua Nikolaus von Schonberg wrote a letter to Copernicus from Rome encouraging him to publish a full version of his theory.

However, in 1539, Martin Luther said:

> *"There is talk of a new astrologer who wants to prove that the earth moves and goes around instead of the sky, the sun, the moon, just as if somebody were moving in a carriage or ship might hold that he was sitting still and at rest while the earth and the trees walked and moved. But that is how things are nowadays: when a man wishes to be clever he must... invent something special, and the way he does it must needs be the best! The fool wants to turn the whole art of astronomy upside-down. However, as Holy Scripture tells us, so did Joshua bid the sun to stand still and not the earth."*

This was reported in the context of a conversation at the dinner table and not a formal statement of faith. Melanchthon, however, opposed the doctrine over a period of years.

Publication of de Revolutionibus (1543)

Nicolaus Copernicus published the definitive statement of his system in *De Revolutionibus* in 1543. Copernicus began to write it in 1506 and finished it in 1530, but did not publish it until the year of his death. Although he was in good standing with the Church and had dedicated the book to Pope Paul III, the published form contained an unsigned preface by Osiander defending the system and arguing that it was useful for computation even if its hypotheses were not necessarily true. Possibly because of that preface, the work of Copernicus inspired very little debate on whether it might be heretical during the next 60 years. There was an early suggestion among Dominicans that the teaching of heliocentrism should be banned, but nothing came of it at the time.

Some years after the publication of *De Revolutionibus* John Calvin preached a sermon in which he denounced those who "pervert the course of nature" by saying that "the sun does not move and that it is the earth that revolves and that it turns". On the other hand, Calvin is not responsible for another famous quotation which has often been misattributed to him:

> *"Who will venture to place the authority of Copernicus above that of the Holy Spirit?"*

It has long been established that this line cannot be found in any of Calvin's works. It has been suggested that the quotation was originally sourced from the works of Lutheran theologian Abraham Calovius.

Tycho Brahe's Geo-heliocentric System c. 1587

Prior to the publication of *De Revolutionibus*, the widely accepted system had been that which was proposed by Ptolemy, in which the Earth was the Centre of the universe and all celestial bodies orbited it.

Tycho Brahe advocated an alternative to the Ptolemaic geocentric system, a geo-heliocentric system now known as the Tychonic system in which the five then known planets orbit the sun, while the sun and the moon orbit the earth. The Jesuit astronomers in Rome were at first unreceptive to Tycho's system; the most prominent, Clavius, commented that Tycho was "confusing all of astronomy, because he wants to have Mars lower than the Sun."

Publication of Starry Messenger (1610)

Galileo was able to look at the night sky with the newly invented telescope. He published his discoveries in Sidereus Nuncius including (among other things) the moons of Jupiter and that Venus exhibited a full range of phases. These discoveries were not consistent with the Ptolemeic model of the solar system. As the Jesuit astronomers confirmed Galileo's observations, the Jesuits moved toward Tycho's teachings.

Publication of Letter to the Grand Duchess (1615)

In a Letter to the Grand Duchess Christina, Galileo defended heliocentrism, and claimed it was not contrary to Scriptures. He took Augustine's position on Scripture: not to take every passage literally when the scripture in question is a book of poetry and songs, not a book of instructions or history. The writers of the Scripture wrote from the perspective of the terrestrial world, and from that vantage point the sun does rise and set. In fact, it is the Earth's rotation which gives the impression of the sun in motion across the sky.

The Decree of 1616

The Letter to the *Grand Duchess Christina* prompted the papal authorities to decide whether heliocentrism was acceptable. Galileo was summoned to Rome to defend his position. The Church accepted the use of heliocentrism as a calculating device, but opposed it as a literal description of the solar system. Cardinal Robert Bellarmine himself considered that Galileo's model made "excellent good sense" on the ground of mathematical simplicity; that is, as a *hypothesis*. And he said:

"If there were a real proof that the Sun is in the Centre of the universe, that the Earth is in the third sphere, and that the Sun does not go round the Earth but the Earth round the Sun, then we should have to proceed with great circumspection in explaining passages of Scripture which appear to teach the contrary, and we should rather have to say that we did not understand them than declare an opinion false which has been proved to be true. But I do not think there is any such proof since none has been shown to me." —Koestler (1959)

Bellarmine supported a ban on the teaching of the idea as anything but hypothesis. In 1616 he delivered to Galileo the papal command not to "hold or defend" the heliocentric idea. The Vatican files suggest that Galileo was forbidden to teach heliocentrism in any way whatsoever, but whether this ban was known to Galileo is a matter of dispute.

Publication of Epitome Astronomia Copernicanae (1617-1621)

In Astronomia nova (1609), Johannes Kepler had used an elliptical orbit to explain the motion of Mars. In *Epitome astronomia Copernicanae* he developed a heliocentric model of the solar system in which all the planets have elliptical orbits. This provided significantly increased accuracy in predicting the position of the planets. Kepler's ideas were not immediately accepted. Galileo for example completely ignored Kepler's work. Kepler proposed heliocentrism as a physical description of the solar system and *Epitome astronomia Copernicanae* was placed on the index of prohibited books despite Kepler being a Protestant.

Publication of Dialogue Concerning the Two Chief World Systems

Pope Urban VIII encouraged Galileo to publish the pros and cons of Heliocentrism. In the event, Galileo's *Dialogue concerning the two chief world systems* clearly advocated heliocentrism and appeared to make fun of the Pope. Urban VIII became hostile to Galileo and he was again summoned to Rome. Galileo's trial in 1633 involved making fine distinctions between "teaching" and "holding and defending as true".

For advancing heliocentric theory Galileo was put under house arrest for the last few years of his life.

According to J. L. Heilbron, Catholic scientists have also:

> *"appreciated that the reference to heresy in connection with Galileo or Copernicus had no general or theological significance."*
> —*Heilbron (1999)*

Subsequent Developments

The Church's opposition to heliocentrism as a literal description did not by any means imply opposition to all astronomy; indeed, it needed observational data to maintain its calendar. In support of this effort it allowed the cathedrals themselves to be used as solar observatories called *meridiane*; i.e., they were turned into "reverse sundials", or gigantic pinhole cameras, where the Sun's image was projected from a hole in a window in the cathedral's lantern onto a meridian line.

In 1664, Pope Alexander VII published his *Index Librorum Prohibitorum Alexandri VII Pontificis Maximi jussu editus* (Index of Prohibited Books, published by order of Alexander VII, P.M.) which included all previous condemnations of heliocentric books.

An annotated copy of Philosophiae Naturalis Principia Mathematica by Isaac Newton was published in 1742 by Fathers le Seur and Jacquier of the Franciscan Minims, two Catholic mathematicians with a preface stating that the author's work assumed heliocentrism and could not be explained without the theory. In 1758 the Catholic Church dropped the general prohibition of books advocating heliocentrism from the *Index of Forbidden Books*. Pope Pius VII approved a decree in 1822 by the Sacred Congregation of the Inquisition to allow the printing of heliocentric books in Rome.

Heliocentrism and Judaism

Already in the Talmud Greek philosophy and science under general name «Greek wisdom» were considered dangerous. It was put under ban then and later for some periods (for example, Shlomo ben Aderet prohibited study of philosophy in 1305). Possibly due to this the system of Nicolaus Copernicus did not cause furious resistance, although it was found to be contradicting verses of Tanakh (Jewish Bible).

The first to mention the new system was Maharal of Prague, although he did not mention Copernicus, the author of the system. In his book "Be'er ha-Golah", 1593 Maharal used the appearance of the new system to show that scientific theories are not reliable enough-even astronomy was turned upside-down.

Copernicus is mentioned for the first time in Hebrew in books of David Gans (1541-1613), who worked with Tycho Brahe and Johannes Kepler. Gans wrote two books on astronomy: a short one "Magen David" (1612) and a full one "Nehmad veNaim" (published only in 1743). He described objectively three systems: Ptolemy, Copernicus and of Tycho Brahe without taking sides.

In 1629 a new Hebrew book "Elim" by Joseph Solomon Delmedigo (1591 – 1655) appeared. The author says that the arguments of Copernicus are so strong, that only an imbecile will not accept them. Delmedigo studied at Padua and was acquainted with Galileo.

The following wave of Hebrew literature on the subject is from the 18th century. Most of the authors were for Copernicus, except David Nieto and Toby Hacohen. The first merely rejected the new system as contradicting to the Bible without much passion. The second one went so far as called Copernicus «a first-born of Satan». The reason is the same, contradiction to the Bible, although Toby mention that Sages of Talmud derived the Hebrew name of Earth from the verb "run".

In later periods there were no explicit attacks on heliocentrism, although some Rabbis were not sure of it.

In the 20th century R. M.M. Schneerson thought that the theory of relativity makes the question obsolete.

The View of Modern Science

Kepler's laws of planetary motion were used as arguments in favor of the heliocentric hypothesis. An apparent proof of the heliocentric hypothesis was provided in 1838 by Friedrich Wilhelm Bessel. Bessel proved that parallax of a star was greater than zero. He measured the parallax of 0.314 arcseconds of a star named 61 Cygni. In the same year Friedrich Georg Wilhelm Struve and Thomas Henderson measured the parallaxes of other stars, Vega and Alpha Centauri.

The thinking that the heliocentric view was also not true in a strict sense was achieved in steps. That the Sun was not the Centre of the universe, but one of innumerable stars, was strongly advocated by the mystic Giordano Bruno. Over the course of the 18th and 19th centuries, the status of the Sun as merely one star among many became increasingly obvious. By the 20th century, even before the discovery that there are many galaxies, it was no longer an issue.

Even if the discussion is limited to the solar system, the sun is not at the geometric Centre of any planet's orbit, but rather at one focus of the elliptical orbit. Furthermore, to the extent that a planet's mass cannot be neglected in comparison to the Sun's mass, the Centre of gravity of the solar system is displaced slightly away from the Centre of the Sun. (The masses of the planets, mostly Jupiter, amount to 0.14% of that of the Sun.) Therefore a hypothetical astronomer on an extrasolar planet would observe a "wobble" in their perception of the Sun's motion.

The concept of an absolute velocity, including being "at rest" as a particular case, is ruled out by the principle of relativity, eliminating any obvious "Centre" of the universe as a natural origin of coordinates. Some forms of Mach's principle consider the frame at rest with respect to the distant masses in the universe to have special properties.

Modern use of Geocentric and Heliocentric

In modern calculations, the origin and orientation of a coordinate system often are selected for practical reasons, and in such systems the origin in the Centre of mass of the Earth, of the Earth-Moon system, of the Sun, of the Sun plus the major planets, or of the entire solar system can be selected. However, such selection of "geocentric" or "heliocentric" coordinates has only practical implications and not philosophical or physical ones.

The View of the Public

A proportion of the public still believes in the geocentric model. Approximately one in five Americans and Britons believe that the Sun revolves around the Earth, according to surveys in 1999, 2006.

Season

Seasons result from the yearly revolution of the Earth around the Sun and the tilt of the Earth's axis relative to the plane of revolution. In temperate and polar regions, the seasons are marked by changes in the intensity of sunlight that reaches the Earth's surface, variations of which may cause animals to go into hibernation or to migrate, and plants to be dormant.

During May, June and July, the northern hemisphere is exposed to more direct sunlight because the hemisphere faces the sun. The same is true of the southern hemisphere in November, December and January. It is the tilt of the Earth that causes the Sun to be higher in the sky during the summer months which increases the solar flux. However, due to seasonal lag, June, July and August are the hottest months in the northern hemisphere and December, January and February are the hottest months in the southern hemisphere.

In temperate and subpolar regions, generally four calendar-based seasons (with their adjectives) are recognized: *spring* (*vernal*), *summer* (*estival*), *autumn* (*autumnal*) and *winter* (*hibernal*). However, ecologists are increasingly using a six-season model for temperate climate regions that includes *pre-spring* (*prevernal*) and *late summer* (*seritonal*) as distinct seasons along with the traditional four.

In some tropical and subtropical regions it is more common to speak of the rainy (or wet, or monsoon) season versus the dry season, because the amount of precipitation may vary more dramatically than the average temperature. For example, in Nicaragua, the dry season (November to April) is called 'summer' and the rainy season (May to October) is called 'winter', even though it is located in the northern hemisphere.

In other tropical areas a three-way division into hot, rainy, and cool season is used.

In some parts of the world, special "seasons" are loosely defined based on important events such as a hurricane season, tornado season or a wildfire season.

Causes and Effects

The seasons result from the Earth's axis being tilted to its orbital plane; it deviates by an angle of approximately 23.4 degrees. Thus, at any given time during summer or winter, one part of the planet is more directly exposed to the rays of the Sun. This exposure alternates as the Earth revolves in its orbit. Therefore, at any given time, regardless of season, the northern and southern hemispheres experience opposite seasons. The effect of axis tilt is observable from the change in day length, and altitude of the Sun at noon (the culmination of the Sun), during a year.

Seasonal weather differences between hemispheres are further caused by the elliptical orbit of Earth. Earth reaches perihelion (the point in its orbit closest to the Sun) in January, and it reaches aphelion (farthest point from the Sun) in July. Even though the effect this has on Earth's seasons is minor, it does noticeably soften the northern hemisphere's winters and summers. In the southern hemisphere, the opposite effect is observed.

Seasonal weather fluctuations (changes) also depend on factors such as proximity to oceans or other large bodies of water, currents in those oceans, El Nino/ENSO and other oceanic cycles, and prevailing winds.

In the temperate and polar regions, seasons are marked by changes in the amount of sunlight, which in turn often causes cycles of dormancy in plants and hibernation in animals. These effects vary with latitude and with proximity to bodies of water. For example, the South Pole is in the middle of the continent of Antarctica and therefore a considerable distance from the moderating influence of the southern oceans. The North Pole is in the Arctic Ocean, and thus its temperature extremes are buffered by the water. The result is that the South Pole is consistently colder during the southern winter than the North Pole during the northern winter. The cycle of seasons in the polar and temperate zones of one hemisphere is opposite to that in the other. When it is summer in the Northern Hemisphere, it is winter in the Southern Hemisphere, and vice versa.

In the tropics, there is no noticeable change in the amount of sunlight. However, many regions (such as the northern Indian ocean) are subject to monsoon rain and wind cycles. A study of temperature records over the past 300 years shows that the climatic seasons, and thus the seasonal year, are governed by the anomalistic year rather than the tropical year.

In meteorological terms, the summer solstice and winter solstice (or the maximum and minimum insolation, respectively) do not fall in the middles of summer and winter. The heights of these seasons occur up to seven weeks later because of seasonal lag. Seasons, though, are not always defined in meteorological terms.

Compared to axial tilt, other factors contribute little to seasonal temperature changes. The seasons are not the result of the variation in Earth's distance to the sun because of its elliptical orbit. Orbital eccentricity can influence temperatures, but on Earth, this effect is small and is more than counteracted by other factors; research shows that the Earth as a whole is actually slightly warmer when *farther* from the sun. This is because the northern hemisphere has more land than the southern, and land warms more readily than sea. Mars however experiences wide temperature variations and violent dust storms every year at perihelion.

Polar Day and Night

Any point north of the Arctic Circle or south of the Antarctic Circle will have one period in the summer when the sun does not set, and one period in the winter when the sun does not rise. At progressively higher latitudes, the maximum periods of "midnight sun" and "polar night" are progressively longer. For example, at the military and weather station Alert on the northern tip of Ellesmere Island, Canada (about 450 nautical miles or 830 km from the North Pole), the sun begins to peek above the horizon in mid-February and each day it climbs higher and stays up longer; by 21 March, the sun is up for 12 hours. However, mid-February is not first light. The sky (as seen from Alert) has twilight, or at least a pre-dawn glow on the horizon, for increasing hours each day, for more than a month before the sun first appears. In the weeks surrounding 21 June, the sun is at its highest, and it appears to circle the sky without going below the horizon. Eventually, it does go below the horizon, for progressively longer periods each day until, around the middle of November, it disappears for the last time. For a few more weeks, "day" is marked by decreasing periods of twilight.

Eventually, for the weeks surrounding 21 December, it is continuously dark. In later winter, the first faint wash of light briefly touches the horizon (for just minutes per day), and then increases in duration and pre-dawn brightness each day until sunrise in February.

Reckoning

Meteorological: Meteorological seasons are reckoned by temperature, with summer being the hottest quarter of the year and winter the coldest quarter of the year. Using this reckoning, the Roman calendar began the year and the spring season on the first of March, with each season occupying three months. In 1780 the Societas Meteorologica Palatina, an early international organization for meteorology, defined seasons as groupings of three whole months. Ever since, professional meteorologists all over the world have used this definition. Therefore, in meteorology for the Northern hemisphere, spring begins on 1 March, summer on 1 June, autumn on 1 September, and winter on 1 December.

In Sweden and Finland, meteorologists use a different definition for the seasons, based on the temperature: spring begins when the daily averaged temperature permanently rises above 0° C, summer begins when the temperature permanently rises above +10° C, summer ends when the temperature permanently falls below +10° C and winter begins when the temperature permanently falls below 0° C. "Permanently" here means that the daily averaged temperature has remained above or below the limit for seven consecutive days. This implies two things: first, the seasons do not begin at fixed dates but must be determined by observation and are known only after the fact; and second, a new season begins at different dates in different parts of the country.

Astronomical

The following diagram shows the relation between the line of solstice and the line of apsides of Earth's elliptical orbit.

The orbital ellipse (with eccentricity exaggerated for effect) goes through each of the six Earth images, which are sequentially the perihelion (periapsis—nearest point to the sun) on anywhere from 2

January to 5 January, the point of March equinox on 20 or 21 March, the point of June solstice on 20 or 21 June, the aphelion (apoapsis—farthest point from the sun) on anywhere from 4 July to 7 July, the September equinox on 22 or 23 September, and the December solstice on 21 or 22 December.

In astronomical reckoning, the solstices and equinoxes ought to be the middle of the respective seasons, but, because of thermal lag, regions with a continental climate often consider these four dates to be the start of the seasons as in the diagram, with the cross-quarter days considered seasonal midpoints. The length of these seasons is not uniform because of the elliptical orbit of the earth and its different speeds along that orbit.

From the March equinox it takes 92.75 days until the June solstice, then 93.65 days until the September equinox, 89.85 days until the December solstice and finally 88.99 days until the March equinox. In Canada and the United States, the mass media consider the astronomical seasons "official" over all other reckonings, but no legal basis exists for this designation.

Because of the differences in the Northern and Southern Hemispheres, it is no longer considered appropriate to use the northern-seasonal designations for the astronomical quarter days. The modern convention for them is: March Equinox, June Solstice, September Equinox and December Solstice. The oceanic climate of the Southern Hemisphere produces a shorter temperature lag, so the start of each season is usually considered to be several weeks before the respective solstice or equinox in this hemisphere, in other countries with oceanic climates, and in cultures with Celtic roots.

Ecological Seasons

Ecologically speaking, a season is a period of the year in which only certain types of floral and animal events happen (e.g.: flowers bloom—spring; hedgehogs hibernate—winter). So, if we can observe a change in daily floral/animal events, the season is changing.

Hot Regions

Hot regions have two seasons:

- Rainy season (winter and spring)
- Dry season (summer and autumn).

Temperate Areas

Six seasons can be distinguished. Mild temperate regions tend to experience the beginning of the hibernal season up to a month later than cool temperate areas, while the prevernal and vernal seasons begin up to a month earlier. For example, prevernal crocus blooms typically appear as early as February in mild coastal areas of British Columbia, the British Isles, and western and southern Europe. The actual dates for each season vary by climate region and can shift from one year to the next. Average dates listed here are for cool temperate climate zones in the Northern Hemisphere:

- Prevernal (ca.1 March–1 May)
- Vernal (ca.1 May–15 June)
- Estival (ca.15 June–15 August)
- Serotinal (ca.15 August–15 September)
- Autumnal (ca.15 September–1 November)
- Hibernal (ca.1 November–1 March).

Cold Regions

There are again only two seasons:

- Polar Day (spring and summer)
- Polar Night (autumn and winter).

Traditional Season Divisions

Traditional seasons are reckoned by insolation, with summer being the quarter of the year with the greatest insolation and winter the quarter with the least. These seasons begin about four weeks earlier than the meteorological seasons and 7 weeks earlier than the astronomical seasons.

In traditional reckoning, the seasons begin at the cross-quarter days. The solstices and equinoxes are the *midpoints* of these seasons. For example, the days of greatest and least insolation are considered the "midsummer" and "midwinter" respectively.

This reckoning is used by various traditional cultures in the Northern Hemisphere, including East Asian and Irish cultures. In Iran, Afghanistan and some other parts of Middle East the beginning of the astronomical spring is the beginning of the new year which is called Nowruz.

So, according to traditional reckoning, winter begins between 5 November and 10 November, Samhain, (lidong or rittou); spring between 2 February and 7 February, Imbolc, (lichun or rissyun); summer between 4 May and 10 May, Beltane (lixia or rikka); and autumn between 3 August and 10 August, Lughnasadh, (liqiu or rissyuu). The middle of each season is considered Mid-winter, between 20 December and 23 December, (dongzhi or touji); Mid-spring, between

19 March and 22 March, (chunfen or syunbun); Mid-summer, between 19 June and 23 June (xiazhi or geshi); and Mid-autumn, between 21 September and 24 September, (qiufen or syuubun).

Australia

The traditional aboriginal people of Australia defined the seasons by what was happening to the plants, animals and weather around them. This led to each separate tribal group having different seasons, some with up to eight seasons each year. However, most modern Aboriginal Australians follow either four or six meteorological seasons, as do non Aboriginal Australians. The commonly followed dates are as follows: 1st day of March, June, September and December for the start of Autumn, Winter, Spring and Summer, respectively.

Celts

The ancient celtic people only recognised two seasons, that of summer and winter. These were marked by festivals, namely Samhain and Beltaine, to celebrate the death and rebirth of the sun.

China

Chinese seasons are traditionally based on 24 periods known as solar terms, and begin at the midpoint of solstices and equinoxes.

India

In India, and in the Hindu calendar, there are six seasons or Ritu: Hemant (pre-winter), Shishir (Winter), Vasanta (Spring), Greeshma (Summer), Varsha (Rainy) and Sharad (Autumn).

Axial Tilt

In astronomy, axial tilt (also called obliquity) is the angle between an object's rotational axis, and a line perpendicular to its orbital plane. It differs from inclination. To measure obliquity, use the right hand grip rule for both the rotation and the orbital motion, i.e.: the line from the vertex at the object's centre to its north pole (above which the object appears to rotate counter-clockwise); and the line drawn from the vertex in the direction of the normal to its orbital plane. At zero degrees, these lines point in the same direction.

The planet Venus has an axial tilt of 177.3° because it is rotating in retrograde direction, opposite to other planets like Earth. The north pole of Venus is pointed 'downward' (our southward). The planet Uranus is rotating on its side in such a way that its rotational axis, and hence its north pole, is pointed almost in the direction of its orbit around the Sun. Hence the axial tilt of Uranus is 97°. Over the course of an orbit, while the angle of the axial tilt doesn't change, the orientation of a planet's axial tilt moves through 360 degrees (one complete orbit around the Sun), relative to the Sun, causing the seasons.

Obliquity

In the Earth's solar system, the Earth's orbital plane is known as the ecliptic plane, and so the Earth's axial tilt is officially called the obliquity of the ecliptic. Greek letter ε. The Earth currently has an axial tilt of about 23.4°. The axis remains tilted in the same direction towards the stars throughout a year and this means that when a hemisphere (a northern or southern half of the earth) is pointing away from the Sun at one point in the orbit then half an orbit later (half a year later) this hemisphere will be pointing towards the Sun. This effect is the main cause of the seasons. Whichever hemisphere is currently tilted toward the Sun experiences more hours of sunlight each day, and the sunlight at midday also strikes the ground at an angle nearer the vertical and thus delivers more energy per unit surface area.

Lower obliquity causes polar regions to receive less seasonally contrasting solar radiation, producing conditions more favourable to glaciation. Like changes in precession and eccentricity, changes in tilt influence the relative strength of the seasons, but the effects of the tilt cycle are particularly pronounced in the high latitudes where the great ice ages began. Obliquity is a major factor in glacial/interglacial fluctuations.

The obliquity of the ecliptic is not a fixed quantity but changing over time in a cycle with a period of 41,000 years. It is a very slow effect known as nutation, and at the level of accuracy at which astronomers work, does need to be taken into account on a daily basis. Note that the obliquity and the precession of the equinoxes are calculated from the same theory and are thus related to each other. A smaller ε means a larger p (precession in longitude) and vice versa. Yet the two movements act independent from each other, going in mutually perpendicular directions.

Measurement

Knowledge of the obliquity of the ecliptic (ε) is critical for astronomical calculations and observations from the surface of the earth (earth-based, positional astronomy).

To quickly grasp an idea of its numerical value one can look at how the sun's angle above the horizon varies with the seasons. The measured difference between the angles of the Sun above the horizon at noon on the longest and shortest days of the year gives twice the obliquity. To an observer on the equator standing all year long looking above, the sun will be directly overhead at noon on the March Equinox, then swing north until it is over the Tropic of Cancer, 23° 26' away from the equator on the Northern Solstice. On the September Equinox it will be back overhead, then swing south until it is over the Tropic of Capricorn, 23° 26' away from the equator on the Southern Solstice.

Example: an observer at 50° latitude (either north or south) will see the Sun 63° 26' above the horizon at noon on the longest day of the year, but only 16° 34' on the shortest day. The difference is 2ε = 46° 52', and so ε = 23° 26'.

(90°- 50°) + 23.4394° = 63.4394° when measuring angles from the horizon (90° - 50°) - 23.4394° = 16.5606°

At the Equator, this would be 90° + 23.4394° = 113.4394° and 90° - 23.4394° = 66.5606° (measuring always from the southern horizon).

Values

The Earth's axial tilt varies between 22.1° and 24.5°, with a 42,000 year period, and at present, the tilt is decreasing. In addition to this steady decrease there are much smaller short term (18.6 years) variations, known as nutation, mainly due to the changing plane of the moon's orbit. This can shift the Earth's axial tilt by plus or minus 0.005 degree.

Simon Newcomb's calculation at the end of the nineteenth century for the obliquity of the ecliptic gave a value of 23° 27' 8.26" (epoch of 1900), and this was generally accepted until improved telescopes allowed more accurate observations, and electronic computers permitted more elaborate models to be calculated. Lieske came with an updated model in 1976 with ε equal to 23° 26' 21.448" (epoch of 2000), which is part of the approximation formula recommended by the International Astronomical Union in 2000:

$\varepsilon = 84381.448 - 46.84024T - (59 \times 10^{5})T^{2} + (1.813 \times 10^{-3})T^{3}$, measured in seconds of arc, with T being the time in Julian centuries (that is, 365,25 days) since the ephemeris epoch of 2000 (which occurred on Julian day 2,451,545.0). A straight application of this formula to 1900 (T=-1) returns 23° 27' 8.29", which is very close to Newcomb's value.

With the linear term in T being negative, at present the obliquity is slowly decreasing. It is implicit that this expression gives only an approximate value for ε and is only valid for a certain range of values of T. If not, ε would approach infinity as T approaches infinity.

Computations based on a numerical model of solar system show that ε has a period of about 41,000 years, the same as the constants of the precession p of the equinoxes (although not of the precession itself).

Other theoretical models may come with values for ε expressed with higher powers of *T*, but since no (finite) polynomial can ever represent a periodic function, they all go to either positive or negative infinity for large enough *T*. In that respect one can understand the decision of the International Astronomical Union to choose the simplest equation which agrees with most models. For up to 5,000 years in the past and the future all formulas agree, and up to 9,000 years in the past and the future, most agree to reasonable accuracy. For eras farther out discrepancies get too large.

Long Period Variations

Nevertheless extrapolation of the average polynomials gives a fit to a sine curve with a period of 41,013 years, which, according to Wittmann, is equal to:

$\varepsilon = A + B \sin(C(T + D))$; with $A = 23.496932° \pm 0.001200°$, $B =$ " $0.860° \pm 0.005°$, $C = 0.01532 \pm 0.0009$ radians/Julian century, $D = 4.40 \pm 0.10$ Julian centuries, and *T*, the time in centuries from the epoch of 2000 as above.

This means a range of the obliquity from 22° 38' to 24° 21', the last maximum was reached in 8700 BC, the mean value occurred around 1550 and the next minimum will be in 11800. This formula should give a reasonable approximation for the previous and next million years or so. Yet it remains an approximation in which the amplitude of the wave remains the same, while in reality, as seen from the results of the Milankovitch cycles, irregular variations occur. The quoted range for the obliquity is from 21° 30' to 24° 30', but the low value may have been a one-time overshot of the normal 22° 30'.

Over the last 5 million years, the obliquity of the ecliptic (or more accurately, the obliquity of the Equator on the moving ecliptic of date) has varied from 22.0425° to 24.5044°, but for the next one million years, the range will be only from 22.2289° to 24.3472°.

Other planets may have a variable obliquity, too; for example, on Mars, the range is believed to be between 11° and 49° as a result of gravitational perturbations from other planets. The relatively small range for the Earth is due to the stabilizing influence of the Moon, but it will not remain so. According to W.R. Ward, the orbit of the Moon (which is continuously increasing due to tidal effects) will have gone from the current 60 to approximately 66.5 Earth radii in about 1.5

billion years. Once this occurs, a resonance from planetary effects will follow, causing swings of the obliquity between 22° and 38°. Further, in approximately 2 billion years, when the Moon reaches a distance of 68 Earth radii, another resonance will cause even greater oscillations, between 27° and 60°. This would have extreme effects on climate.

Biosphere

Our biosphere is the global sum of all ecosystems. It can also be called the zone of life on Earth, a closed (apart from solar and cosmic radiation) and self-regulating system. From the broadest biophysiological point of view, the biosphere is the global ecological system integrating all living beings and their relationships, including their interaction with the elements of the lithosphere, hydrosphere and atmosphere. The biosphere is postulated to have evolved, beginning through a process of biogenesis or biopoesis, at least some 3.5 billion years ago.

In a broader sense; biospheres are any closed, self-regulating systems containing ecosystems; including artificial ones such as Biosphere 2 and BIOS-3; and, potentially, ones on other planets or moons.

Origin and Use of the Term

The term "biosphere" was coined by geologist Eduard Suess in 1875, which he defined as:

"The place on Earth's surface where life dwells."

While this concept has a geological origin, it is an indication of the impact of both Darwin and Maury on the earth sciences. The biosphere's ecological context comes from the 1920s, preceding the 1935 introduction of the term "ecosystem" by Sir Arthur Tansley. Vernadsky defined ecology as the science of the biosphere. It is an interdisciplinary concept for integrating astronomy, geophysics, meteorology, biogeography, evolution, geology, geochemistry, hydrology and, generally speaking, all life and earth sciences.

Gaia Hypothesis

The concept that the biosphere is itself a living organism, either actually or metaphorically, is known as the Gaia hypothesis.

James Lovelock, an atmospheric scientist from the United Kingdom, proposed the Gaia hypothesis to explain how biotic and abiotic factors interact in the biosphere. This hypothesis considers Earth itself a kind of living organism. Its atmosphere, geosphere, and hydrosphere are cooperating systems that yield a biosphere full of life. In the early 1970s, Lynn Margulis, a microbiologist from the United States, added to the hypothesis, specifically noting the ties between the biosphere and other Earth systems. For example, when carbon dioxide levels increase in the atmosphere, plants grow more quickly. As their growth continues, they remove more and more carbon dioxide from the atmosphere.

Many scientists are now involved in new fields of study that examine interactions between biotic and abiotic factors in the biosphere, such as geobiology and geomicrobiology.

Ecosystems occur when communities and their physical environment work together as a system. The difference between this and a biosphere is simple, the biosphere is everything in general terms.

Extent of Earth's Biosphere

Every part of the planet, from the polar ice caps to the Equator, supports life of some kind. Recent advances in microbiology have demonstrated that microbes live deep beneath the Earth's terrestrial surface, and that the total mass of microbial life in so-called "uninhabitable zones" may, in biomass, exceed all animal and plant life on the surface. The actual thickness of the biosphere on earth is difficult to measure. Birds typically fly at altitudes of 650 to 1,800 meters, and fish that live deep underwater can be found down to -8,372 meters in the Puerto Rico Trench.

There are more extreme examples for life on the planet: Ruppell's Vulture has been found at altitudes of 11,300 meters; Bar-headed Geese migrate at altitudes of at least 8,300 meters (over Mount Everest); Yaks live at elevations between 3,200 to 5,400 meters above sea level; mountain goats live up to 3,050 meters. Herbivorous animals at these elevations depend on lichens, grasses, and herbs.

Microscopic organisms live at such extremes that, taking them into consideration puts the thickness of the biosphere much greater. Culturable microbes have been found in the Earth's upper atmosphere as high as 41 km (25 mi) (Wainwright et al., 2003, in FEMS Microbiology Letters). It is unlikely, however, that microbes are active at such altitudes, where temperatures and air pressure are extremely low and ultraviolet radiation very high. More likely these microbes were brought into the upper atmosphere by winds or possibly volcanic eruptions. Barophilic marine microbes have been found at more than 10 km (6 mi) depth in the Marianas Trench (Takamia et al., 1997, in FEMS Microbiology Letters).

Microbes are not limited to the air, water or the Earth's surface. Culturable thermophilic microbes have been extracted from cores drilled more than 5 km (3 mi) into the Earth's crust in Sweden, from rocks between 65-75 °C. Temperature increases with increasing depth into the Earth's crust. The speed at which the temperature increases depends on many factors, including type of crust (continental vs. oceanic), rock type, geographic location, etc. The upper known limit of microbial is 122 °C, and it is likely that the limit of life in the "deep biosphere" is defined by temperature rather than absolute depth.

Our biosphere is divided into a number of biomes, inhabited by broadly similar flora and fauna. On land, biomes are separated primarily by latitude. Terrestrial biomes lying within the Arctic and Antarctic Circles are relatively barren of plant and animal life, while most of the more populous biomes lie near the equator. Terrestrial organisms in temperate and Arctic biomes have relatively small amounts of total biomass, smaller energy budgets, and display prominent adaptations to cold, including world-spanning migrations, social adaptations, homeothermy, estivation and multiple layers of insulation.

Specific Biospheres

When the word is followed by a number, it is usually referring to a specific system or number. Thus:

- Biosphere 1, the planet Earth
- Biosphere 2, a laboratory in Arizona which contains 3.15 acres (13,000 m^2) of closed ecosystem.
- BIOS-3, a closed ecosystem at the Institute of Biophysics in Krasnoyarsk, Siberia, in what was then the Soviet Union.
- Biosphere J (CEEF, Closed Ecology Experiment Facilities), a experiment in Japan.

Natural Resource

Natural resources (economically referred to as land or raw materials) occur naturally within environments that exist relatively undisturbed by mankind, in a natural form. A natural resource is often characterized by amounts of biodiversity existent in various ecosystems. Natural resources are derived from the environment. This is currently restricted to the environment of Earth yet the theoretical possibility remains of extracting them from outside the planet, such as the asteroid belt. Many of them are essential for our survival while others are used for satisfying our wants. Natural resources may be further classified in different ways.

Classification

On the basis of origin, resources may be divided into:

- *Biotic*-Biotic resources are obtained from the biosphere, such as forests and their products, animals, birds and their products, fish and other marine organisms. Mineral fuels such as coal and petroleum are also included in this category because they are formed from decayed organic matter.
- *Abiotic*-Abiotic resources include non-living things. Examples include land, water, air and ores such as gold, iron, copper, silver etc.

Considering their stage of development, natural resources may be referred to in the following ways:

- *Potential Resources*-Potential resources are those that exist in a region and may be used in the future. For example, petroleum may exist in many parts of India, having sedimentary rocks but until the time it is actually drilled out and put into use, it remains a potential resource.
- *Actual Resources* are those that have been surveyed, their quantity and quality determined and are being used in present times. The development of an actual resource, such as wood processing depends upon the technology available and the cost involved. That part of the actual resource that can be developed profitably with available technology is called a reserve.

On the basis of status of development, they can be classified into potential resources, developed resources, stock and reserves.

With respect to renewability, natural resources can be categorized as follows:

- Renewable resources are ones that can be replenished or reproduced easily. Some of them, like sunlight, air, wind, etc., are continuously available and their quantity is not affected by human consumption. Many renewable resources can be depleted by human use, but may also be replenished, thus maintaining a flow. Some of these, like agricultural crops, take a short time for renewal; others, like water, take a comparatively longer time, while still others, like forests, take even longer.
- Non-renewable resources are formed over very long geological periods. Minerals and fossil fuels are included in this category. Since their rate of formation is extremely slow, they cannot be replenished once they get depleted. Of these, the metallic minerals can be re-used by recycling them. But coal and petroleum cannot be recycled.

On the basis of availability, natural resources can be categorised as follows:

- Inexhaustible natural resources-Those resources which are present in unlimited quantity in nature and are not likely to be exhausted easily by human activity are inexhaustible natural resources (sunlight, air etc.)
- Exhaustible natural resources-The amount of these resources are limited. They can be exhausted by human activity in the long run (coal, petroleum, natural gas, etc.)

Examples

Some examples of natural resources include the following:

- Air, wind and atmosphere
- Plants
- Animals
- Coal, fossil fuels, rock and mineral resources
- Forestry
- Range and pasture
- Soils
- Water, oceans, lakes, groundwater and rivers
- Sun (solar power].

Management

Natural resource management is a discipline in the management of natural resources such as land, water, soil, plants and animals, with a particular focus on how management affects the quality of life for both present and future generations. Natural resource management is interrelated with the concept of sustainable development, a principle that forms a basis for land management and environmental governance throughout the world. In contrast to the policy emphases of urban planning and the broader concept of environmental management, Natural resource management specifically focuses on a scientific and technical understanding of resources and ecology and the life-supporting capacity of those resources.

Depletion

In recent years, the depletion of natural resources and attempts to move to sustainable development has been a major focus of development agencies. This is a particular concern in rainforest regions, which hold most of the Earth's natural biodiversity-irreplaceable genetic natural capital. Conservation of natural resources is the major focus of natural capitalism, environmentalism, the ecology movement, and green politics. Some view this depletion as a major source of social unrest and conflicts in developing nations. Mining, petroleum extraction, fishing, hunting, and forestry are generally considered natural-resource industries. Agriculture is considered a man-made resource. Theodore Roosevelt, a well-known conservationist and former United States president, was opposed to unregulated natural resource extraction. The term is defined by the United States Geological Survey as "The Nation's natural resources include its minerals, energy, land, water, and biota."

Protection

Conservation biology is the scientific study of the nature and status of Earth's biodiversity with the aim of protecting species, their habitats, and ecosystems from excessive rates of extinction. It is an interdisciplinary subject drawing on sciences, economics, and the practice of natural resource management. The term *conservation biology* was introduced as the title of a conference held University of California at San Diego in La Jolla, California in 1978 organized by biologists Bruce Wilcox and Michael Soule.

Habitat conservation is a land management practice that seeks to conserve, protect and restore, habitat areas for wild plants and animals, especially conservation reliant species, and prevent their extinction, fragmentation or reduction in range.

Natural and Environmental Hazards

Large areas of the Earth's surface are subject to extreme weather such as tropical cyclones, hurricanes, or typhoons that dominate life in those areas. Many places are subject to earthquakes, landslides, tsunamis, volcanic eruptions, tornadoes, sinkholes, blizzards, floods, droughts, and other calamities and disasters. Many localized areas are subject to human-made pollution of the air and water, acid rain and toxic substances, loss of vegetation (overgrazing, deforestation, desertification), loss of wildlife, species extinction, soil degradation, soil depletion, erosion, and introduction of invasive species. According to the United Nations, a scientific consensus exists linking human activities to global warming due to industrial carbon dioxide emissions. This is predicted to produce changes such as the melting of glaciers and ice sheets, more extreme temperature ranges, significant changes in weather and a global rise in average sea levels.

Atmosphere of Earth

The atmosphere of Earth is a layer of gases surrounding the planet Earth that is retained by Earth's gravity. The atmosphere protects life on Earth by absorbing ultraviolet solar radiation, warming the surface through heat retention (greenhouse effect), and reducing temperature extremes between day and night.

Atmospheric stratification describes the structure of the atmosphere, dividing it into distinct layers, each with specific characteristics such as temperature or composition. The atmosphere has a mass of about 5×10^{18} kg, three quarters of which is within about 11 km (6.8 mi; 36,000 ft) of the surface. The atmosphere becomes thinner and thinner with increasing altitude, with no definite boundary between the atmosphere and outer space. An altitude of 120 km (75 mi) is where atmospheric effects become noticeable during atmospheric reentry of spacecraft. The Karman line, at 100 km (62 mi), also is often regarded as the boundary between atmosphere and outer space. Air is the name given to atmosphere used in breathing and photosynthesis. Dry air contains roughly (by volume) 78.09% nitrogen, 20.95% oxygen, 0.93% argon, 0.039% carbon dioxide, and small amounts of other gases. Air also contains a variable amount of water vapour, on average around 1%. While air content and atmospheric pressure varies at different layers, air suitable for the survival of terrestrial plants and terrestrial animals is currently known only to be found in Earth's troposphere and artificial atmospheres.

Composition

Air is mainly composed of nitrogen, oxygen, and argon, which together constitute the major gases of the atmosphere. The remaining gases are often referred to as trace gases, among which are the greenhouse gases such as water vapor, carbon dioxide, methane, nitrous oxide, and ozone. Filtered air includes trace amounts of many other chemical compounds. Many natural substances may be present in tiny amounts in an unfiltered air sample, including dust, pollen and spores, sea spray, and volcanic ash. Various industrial pollutants also may be present, such as chlorine (elementary or in compounds), fluorine compounds, elemental mercury, and sulfur compounds such as sulfur dioxide [SO_2].

***Table:** Composition of dry atmosphere, by volume*

ppmv: parts per million by volume (note: volume fraction is equal to mole fraction for ideal gas only)

Gas	*Volume*
Nitrogen (N_2)	780,840 ppmv (78.084%)
Oxygen (O_2)	209,460 ppmv (20.946%)
Argon (Ar)	9,340 ppmv (0.9340%)
Carbon dioxide (CO_2)	390 ppmv (0.039%)
Neon (Ne)	18.18 ppmv (0.001818%)
Helium (He)	5.24 ppmv (0.000524%)
Methane (CH_4)	1.79 ppmv (0.000179%)

Contd...

Gas	Volume
Krypton (Kr)	1.14 ppmv (0.000114%)
Hydrogen (H_2)	0.55 ppmv (0.000055%)
Nitrous oxide (N_2O)	0.3 ppmv (0.00003%)
Carbon monoxide (CO)	0.1 ppmv (0.00001%)
Xenon (Xe)	0.09 ppmv (9×10-6%) (0.000009%)
Ozone (O_3)	0.0 to 0.07 ppmv (0 to 7×10-6%)
Nitrogen dioxide (NO_2)	0.02 ppmv (2×10-6%) (0.000002%)
Iodine (I_2)	0.01 ppmv (1×10-6%) (0.000001%)
Ammonia (NH_3)	trace
Not included in above dry atmosphere:	
Water vapor (H_2O) 1%-4% at surface.	~0.40% over full atmosphere, typically

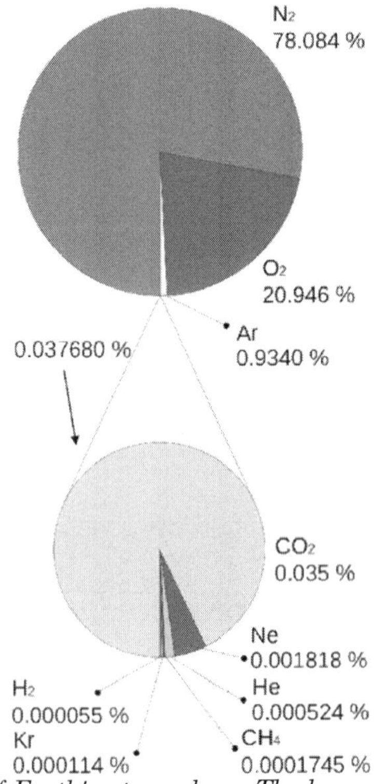

Figure: *Composition of Earth's atmosphere. The lower pie represents the trace gases which together compose **0.039%** of the atmosphere. Values normalized for illustration. The numbers are from a variety of years (mainly 1987, with CO_2 and methane from 2009) and do not represent any single source.*

Structure of the Atmosphere

Principal Layers

In general, air pressure and density decrease in the atmosphere as height increases. However, temperature has a more complicated profile with altitude. Because the general pattern of this profile is constant and recognizable through means such as balloon soundings, temperature provides a useful metric to distinguish between atmospheric layers.

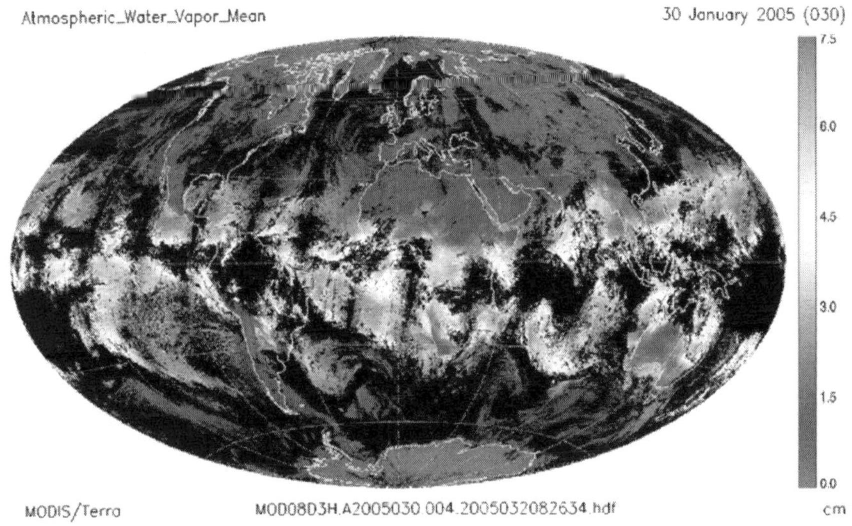

Figure: *Mean atmospheric water vapor*

In this way, Earth's atmosphere can be divided into five main layers. From highest to lowest, these layers are:

Exosphere: The exosphere is the uppermost layer of the atmosphere. In the exosphere, an upward travelling molecule moving fast enough to attain escape velocity can escape to space with a low chance of collisions; if it is moving below escape velocity it will be prevented from escaping from the celestial body by gravity. In either case, such a molecule is unlikely to collide with another molecule due to the exosphere's low density.

Earth's Exosphere: The main gases within the Earth's exosphere are the lightest gases, mainly hydrogen, with some helium, carbon dioxide, and atomic oxygen near the exobase. The exosphere is the last layer before outer space. Since there is no clear boundary between outer space and the exosphere, the exosphere is sometimes considered a part of outer space.

Lower Boundary: The altitude of its lower boundary, known as the *thermopause* and *exobase*, ranges from about 250 to 500 kilometres (160 to 310 mi) depending on solar activity. Its lower boundary at the edge of the thermosphere has sometimes been estimated to be 500 to 1,000 km (310 to 620 mi) above the Earth's surface. The exobase is also called the critical level, the lowest altitude of the exosphere, and is typically defined in one of two ways:

1. The height above which there are the negligible atomic collisions between the particles and
2. The height above which constituent atoms are on purely ballistic trajectories.

Thermosphere: Earth atmosphere diagram showing the exosphere and other layers. The layers are to scale. From Earth's surface to the top of the stratosphere (50 kilometres (31 mi)) is just under 1% of Earth's radius.

The thermosphere is the biggest of all the layers of the earth's atmosphere directly above the mesosphere and directly below the exosphere. Within this layer, ultraviolet radiation causes ionization. The International Space Station has a stable orbit within the middle of the thermosphere, between 320 and 380 kilometres (200 and 240 mi). Auroras also occur in the thermosphere.

Named from the Greek (*thermos*) for heat, the thermosphere begins about 80 kilometres (50 mi) above the earth. At these high altitudes, the residual atmospheric gases sort into strata according to molecular mass. Thermospheric temperatures increase with altitude due to absorption of highly energetic solar radiation by the small amount of residual oxygen still present. Temperatures are highly dependent on solar activity, and can rise to 1,500 °C (2,730 °F). Radiation causes the atmosphere particles in this layer to become electrically charged, enabling radio waves to bounce off and be received beyond the horizon. At the exosphere, beginning at 500 to 1,000 kilometres (310 to 620 mi) above the Earth's surface, the atmosphere turns into space.

The highly diluted gas in this layer can reach 2,500 °C (4,530 °F) during the day. Even though the temperature is so high, one would not feel warm in the thermosphere, because it is so near vacuum that there is not enough contact with the few atoms of gas to transfer much heat. A normal thermometer would read significantly below 0 °C (32 °F), due to the energy lost by thermal radiation overtaking the energy acquired from the atmospheric gas by direct contact. Above 160 kilometres (99 mi), the anacoustic zone prevents the transmission of sound.

The dynamics of the lower thermosphere are dominated by atmospheric tide, which is driven, in part, by the very significant diurnal heating. The atmospheric tide dissipates above this level since molecular concentrations do not support the coherent motion needed for fluid flow.

Mesosphere: The mesosphere (from the Greek words *mesos* = middle and *sphaira* = ball) is the layer of the Earth's atmosphere that is directly above the stratosphere and directly below the thermosphere. The mesosphere is located about 50 to 85 kilometres (30 to 50 miles) above the Earth's surface.

The stratosphere and mesosphere are referred to as the middle atmosphere. The mesopause, at an altitude of 80–90 km (50–56 mi), separates the mesosphere from the thermosphere—the second-outermost layer of the Earth's atmosphere. This is also around the same altitude as the turbopause, below which different chemical species are well mixed due to turbulent eddies. Above this level the atmosphere becomes non-uniform; the scale heights of different chemical species differ by their molecular masses.

Temperature: Within the mesosphere, temperature decreases with increasing altitude. This is due to decreasing solar heating and increasing cooling by CO_2 radiative emission. The top of the mesosphere, called the mesopause, is the coldest place on Earth. Temperatures in the upper mesosphere fall as low as –100 °C (173 K; –148 °F), varying according to latitude and season.

Dynamical Features: The main dynamical features in this region are atmospheric tides, internal atmospheric gravity waves (commonly called "gravity waves") and planetary waves. Most of these tides and waves are excited in the troposphere and lower stratosphere, and propagate upward to the mesosphere. In the mesosphere, gravity-wave amplitudes can become so large that the waves become unstable and dissipate. This dissipation deposits momentum into the mesosphere and largely drives global circulation.

Noctilucent clouds are located in the mesosphere. The mesosphere is also the region of the ionosphere known as the *D layer*. The D layer is only present during the day, when some ionization occurs with nitric oxide being ionized by Lyman series-alpha hydrogen radiation. The ionization is so weak that when night falls, and the source of ionization is removed, the free electron and ion form back into a neutral molecule.

A 5 km (3.1 mi) deep sodium layer is located between 80–105 km (50–65 mi). Made of unbound, non-ionized atoms of sodium, the sodium layer radiates weakly to contribute to the airglow.

Uncertainties: The mesosphere lies above the maximum altitude for aircraft and below the minimum altitude for orbital spacecraft. It has only been accessed through the use of sounding rockets. As a result, it is the most poorly understood part of the atmosphere. The presence of red sprites and blue jets (electrical discharges or lightning within the lower mesosphere), noctilucent clouds and density shears within the poorly understood layer are of current scientific interest.

Meteors: Millions of meteors enter the atmosphere, an average of 40 tons per day. Within the mesosphere most melt or vaporize as a result of collisions with the gas particles contained there. This results in a higher concentration of iron and other refractory materials reaching the surface.

Stratosphere: The stratosphere is the second major layer of Earth's atmosphere, just above the troposphere, and below the mesosphere. It is stratified in temperature, with warmer layers higher up and cooler layers farther down. This is in contrast to the troposphere near the Earth's surface, which is cooler higher up and warmer farther down. The border of the troposphere and stratosphere, the tropopause, is marked by where this inversion begins, which in terms of atmospheric thermodynamics is the equilibrium level. The stratosphere is situated between about 10 km (6 miles) and 50 km (31 miles) altitude above the surface at moderate latitudes, while at the poles it starts at about 8 km (5 miles) altitude.

Ozone and Temperature: Within this layer, temperature increases as altitude increases; the top of the stratosphere has a temperature of about 270 K ("3°C or 29.6°F), just slightly below the freezing point of water. The stratosphere is layered in temperature because ozone (O_3) here absorbs high energy UVB and UVC energy waves from the Sun and is broken down into monoatomic oxygen (O) and diatomic oxygen (O_2). Monoatomic oxygen is found prevalent in the upper stratosphere due to the bombardment of UV light and the destruction of both ozone and diatomic oxygen. The mid stratosphere has less UV light passing through it, O and O_2 are able to combine, and is where the majority of natural ozone is produced. It is when these two forms of oxygen recombine to form ozone that they release the heat found in the stratosphere. The lower stratosphere receives very low amounts of UVC, thus monoatomic oxygen is not found here and ozone is not formed (with heat as the byproduct). This vertical stratification, with warmer layers above and cooler layers below, makes the stratosphere dynamically stable: there is no regular convection and associated turbulence in this part of the atmosphere. The top of the stratosphere is called the stratopause, above which the temperature decreases with height.

Methane (CH_4) while it is not a direct cause of ozone destruction in the stratosphere, does lead to the formation of compounds that do destroy ozone. Monoatomic oxygen (O), in the upper stratosphere, reacts with methane (CH_4) to form a hydroxyl anion (OH). This hydroxyl anion is then able to interact with non-soluble compounds like chlorofluorocarbons and UV light break off chlorine anions (Cl^-). These chlorine anions break off an oxygen atom from the ozone molecule, creating an oxygen molecule (O_2) and a hypochlorite molecule (ClO^-). The hypochlorite molecule then reacts with a monoatomic oxygen creating another oxygen molecule and another chlorine anion, thereby preventing the reaction of a monoatomic oxygen with O_2 to create natural ozone.

Aircraft Flight: Commercial airliners typically cruise at altitudes of 9–12 km (29,000 to 39,000 ft) in temperate latitudes (in the lower reaches of the stratosphere). They do this to optimize fuel burn, mostly thanks to the low temperatures encountered near the tropopause and the low air density that reduces parasitic drag on the airframe. It also allows them to stay above any hard weather (extreme turbulence).

Because the temperature in the tropopause and lower stratosphere remains constant (or slightly increases) with increasing altitude, there is very little convective turbulence at these altitudes. Though most of the turbulence at this altitude is caused by variations in the jet stream and other local wind shears, areas of significant convective activity (thunderstorms) in the troposphere below may produce convective overshoot. Although a few gliders have achieved great altitudes in the powerful thermals in thunderstorms, this is dangerous. Most high altitude flights by gliders use lee waves from mountain ranges and were used to set the current record of 15,447m (50,671 feet).

Circulation and Mixing: The stratosphere is a region of intense interactions among radiative, dynamical, and chemical processes, in which horizontal mixing of gaseous components proceeds much more rapidly than vertical mixing. An interesting feature of stratospheric circulation is the quasi-Biennial Oscillation (QBO) in the tropical latitudes, which is driven by gravity waves that are convectively generated in the troposphere. The QBO induces a secondary circulation that is important for the global stratospheric transport of tracers such as ozone or water vapour.

In northern hemispheric winter, sudden stratospheric warmings can often be observed which are caused by the absorption of Rossby waves in the stratosphere.

Life: Bacterial life survives in the stratosphere, making it a part of the biosphere. Also, some bird species have been reported to fly at the lower levels of the stratosphere. On November 29, 1975, a Ruppell's Vulture was reportedly ingested into a jet engine 37,900 feet above the Ivory Coast, and Bar-headed geese routinely overfly Mount Everest's summit, which is 29,028 feet.

Troposphere: The troposphere is the lowest portion of Earth's atmosphere. It contains approximately 75% of the atmosphere's mass and 99% of its water vapour and aerosols.

The average depth of the troposphere is approximately 17 km (11 mi) in the middle latitudes. It is deeper in the tropical regions, up to 20 km (12 mi), and shallower near the poles, at 7 km (4.3 mi) in summer, and indistinct in winter. The lowest part of the troposphere, where friction with the Earth's surface influences air flow, is the planetary boundary layer. This layer is typically a few hundred meters to 2 km (1.2 mi) deep depending on the landform and time of day. The border between the troposphere and stratosphere, called the tropopause, is a temperature inversion.

The word troposphere derives from the Greek: *tropos* for "turning" or "mixing," reflecting the fact that turbulent mixing plays an important role in the troposphere's structure and behaviour. Most of the phenomena we associate with day-to-day weather occur in the troposphere.

Composition: The chemical composition of the troposphere is essentially uniform, with the notable exception of water vapour. The source of water vapour is at the surface through the processes of evaporation and transpiration. Furthermore the temperature of the troposphere decreases with height, and saturation vapour pressure decreases strongly as temperature drops, so the amount of water vapour that can exist in the atmosphere decreases strongly with height. Thus the proportion of water vapour is normally greatest near the surface and decreases with height.

Pressure: The pressure of the atmosphere is maximum at sea level and decreases with higher altitude. This is because the atmosphere is very nearly in hydrostatic equilibrium, so that the pressure is equal to the weight of air above a given point. Where:

- g_n stands for the standard gravity
- ρ stands for density
- z stands for height
- p stands for pressure

- R stands for the gas constant
- T stands for temperature in kelvins
- m stands for the molar mass.

Since temperature in principle also depends on altitude, one needs a second equation to determine the pressure as a function of height, as discussed in the next section.*

Lapse Rate: The lapse rate is defined as the rate of decrease with height for an atmospheric variable. The variable involved is temperature unless specified otherwise. The terminology arises from the word *lapse* in the sense of a decrease or decline; thus, the lapse rate is the rate of decrease with height and not simply the rate of change. While most often applied to Earth's atmosphere the concept can be extended to any gravitationally supported ball of gas.

Definition: A formal definition from the *Glossary of Meteorology* is:

The decrease of an atmospheric variable with height, the variable being temperature unless otherwise specified.

The term applies ambiguously to the environmental lapse rate and the process lapse rate, and the meaning must often be ascertained from the context.

Types of Lapse Rates: There are two types of lapse rate:

- Environmental lapse rate – which refers to the actual change of temperature with altitude for the stationary atmosphere (i.e. the temperature gradient)
- The adiabatic lapse rates – which refer to the change in temperature of a mass of air as it moves upwards. There are two adiabatic rates:
  - Dry adiabatic lapse rate
  - Moist (or saturated) adiabatic lapse rate.

Environmental Lapse Rate: The environmental lapse rate (ELR), is the rate of decrease of temperature with altitude in the stationary atmosphere at a given time and location. As an average, the International Civil Aviation Organization (ICAO) defines an international standard atmosphere (ISA) with a temperature lapse rate of 6.49 K(°C)/1,000 m (3.56 °F or 1.98 K(°C)/1,000 Ft) from sea level to 11 km (36,090 ft). From 11 km (36,090 ft or 6.8 mi) up to 20 km (65,620 ft or 12.4 mi), the constant temperature is “56.5 °C (“69.7 °F), which is the lowest assumed temperature in the ISA. The standard atmosphere contains no moisture. Unlike the idealized ISA, the temperature of the actual atmosphere does not always fall at a uniform rate with height.

For example, there can be an inversion layer in which the temperature increases with height.

Dry Adiabatic Lapse Rate: The dry adiabatic lapse rate (DALR) is the rate of temperature decrease with height for a parcel of dry or unsaturated air rising under adiabatic conditions. Unsaturated air has less than 100% relative humidity; i.e. its actual temperature is higher than its dew point. The term *adiabatic* means that no heat transfer occurs into or out of the parcel. Air has low thermal conductivity, and the bodies of air involved are very large, so transfer of heat by conduction is negligibly small.

Under these conditions when the air rises (for instance, by convection) it expands, because the pressure is lower at higher altitudes. As the air parcel expands, it pushes on the air around it, doing work. Since the parcel does work but gains no heat, it loses internal energy so that its temperature decreases. The rate of temperature decrease is 9.8 °C per 1,000 m. (The reverse occurs for a sinking parcel of air.)

For an adiabatic process, first law of thermodynamics can be written as :

$$nc_v dT - VdT/\gamma = 0$$

For adiabatic process:

$$PdV = -VdP/\gamma$$

Also since :$\alpha = V/n$ and :$\gamma = c_p/c_v$ we can show that:

$$c_p dT \square - \alpha dP = 0$$

Where c_p is the specific heat at constant pressure and α is the specific volume.

Assuming an atmosphere in hydrostatic equilibrium:

$$dP = -\rho g dz$$

Where g is the standard gravity and ρ is the density. Combining these two equations to eliminate the pressure, one arrives at the result for the DALR.

Saturated Adiabatic Lapse Rate: When the air is saturated with water vapour (at its dew point), the moist adiabatic lapse rate (MALR) or saturated adiabatic lapse rate (SALR) applies. This lapse rate varies strongly with temperature. A typical value is around 5 °C/km (2.7 °F/ 1,000 ft).

The reason for the difference between the dry and moist adiabatic lapse rate values is that latent heat is released when water condenses, thus decreasing the rate of temperature drop as altitude increases.

This heat release process is an important source of energy in the development of thunderstorms. An unsaturated parcel of air of given temperature, altitude and moisture content below that of the corresponding dewpoint cools at the *dry adiabatic lapse rate* as altitude increases until the dewpoint line for the given moisture content is intersected. As the water vapour then starts condensing the air parcel subsequently cools at the slower *moist adiabatic lapse rate* if the altitude increases further.

$$\Gamma_w = g \frac{1 + \frac{H_v r}{R_{sd} T}}{C_{pd} + \frac{H_v^2 r \in}{R_{sd} T^2}}$$

where:

Γ_w = Wet adiabatic lapse rate, K/m

g = Earth's gravitational acceleration = 9.8076 m/s^2

H_v = Heat of vaporization of water, J/kg

r = The ratio of the mass of water vapour to the mass of dry air, kg/kg

R = The universal gas constant = 8,314 J kmol^{-1} K^{-1}

M = The molecular weight of any specific gas, kg/kmol = 28.964 for dry air and 18.015 for water vapour

R/M = The specific gas constant of a gas, denoted as R_s

R_{sd} = Specific gas constant of dry air = 287 J kg^{-1} K^{-1}

R_s = Specific gas constant of water vapour = 462 J kg^{-1} K^{-1}

ε=The dimensionless ratio of the specific gas constant of dry air to the specific gas constant for water vapour = 0.6220

T = Temperature of the saturated air, K

c_{pd} = The specific heat of dry air at constant pressure, J kg^{-1} K^{-1}

Significance in Meteorology: The varying environmental lapse rates throughout the Earth's atmosphere are of critical importance in meteorology, particularly within the troposphere. They are used to determine if the parcel of rising air will rise high enough for its water to condense to form clouds, and, having formed clouds, whether the air will continue to rise and form bigger shower clouds, and whether these clouds will get even bigger and form cumulonimbus clouds (thunder clouds).

As unsaturated air rises, its temperature drops at the dry adiabatic rate. The dew point also drops (as a result of decreasing air pressure) but much more slowly, typically about "2 °C per 1,000 m. If unsaturated air rises far enough, eventually its temperature will reach its dew point, and condensation will begin to form. This altitude is known as the lifting condensation level (LCL) when mechanical lift is present and the convective condensation level (CCL) absent mechanical lift, in which case, the parcel must be heated from below to its convective temperature. The cloud base will be somewhere within the layer bounded by these parameters.

The difference between the dry adiabatic lapse rate and the rate at which the dew point drops is around 8 °C per 1,000 m. Given a difference in temperature and dew point readings on the ground, one can easily find the LCL by multiplying the difference by 125 m/°C.

If the environmental lapse rate is less than the moist adiabatic lapse rate, the air is absolutely stable — rising air will cool faster than the surrounding air and lose buoyancy. This often happens in the early morning, when the air near the ground has cooled overnight. Cloud formation in stable air is unlikely.

If the environmental lapse rate is between the moist and dry adiabatic lapse rates, the air is conditionally unstable — an unsaturated parcel of air does not have sufficient buoyancy to rise to the LCL or CCL, and it is stable to weak vertical displacements in either direction.

If the parcel is saturated it is unstable and will rise to the LCL or CCL, and either be halted due to an inversion layer of convective inhibition, or if lifting continues, deep, moist convection (DMC) may ensue, as a parcel rises to the level of free convection (LFC), after which it enters the free convective layer (FCL) and usually rises to the equilibrium level (EL).

If the environmental lapse rate is larger than the dry adiabatic lapse rate, it has a superadiabatic lapse rate, the air is absolutely unstable — a parcel of air will gain buoyancy as it rises both below and above the lifting condensation level or convective condensation level. This often happens in the afternoon over many land masses. In these conditions, the likelihood of cumulus clouds, showers or even thunderstorms is increased.

Meteorologists use radiosondes to measure the environmental lapse rate and compare it to the predicted adiabatic lapse rate to forecast the likelihood that air will rise. Charts of the environmental lapse rate are known as thermodynamic diagrams, examples of which include Skew-T log-P diagrams and tephigrams.

The difference in moist adiabatic lapse rate and the dry rate is the cause of foehn wind phenomenon (also known as "Chinook winds" in parts of North America).

Tropopause: The tropopause is the atmospheric boundary between the troposphere and the stratosphere. Going upward from the surface, it is the point where air ceases to cool with height, and becomes almost completely dry. More formally, it is the region of the atmosphere where the environmental lapse rate changes from positive (in the troposphere) to negative (in the stratosphere). The exact definition used by the World Meteorological Organization is:

The lowest level at which the lapse rate decreases to 2 °C/km or less, provided that the average lapse rate between this level and all higher levels within 2 km does not exceed 2 °C/km.

The troposphere is the lowest of the Earth's atmospheric layers and is the layer in which most weather occurs. The troposphere begins at ground level and ranges in height from an average of 11 km (6.8 miles/36,080 feet at the International Standard Atmosphere) at the poles to 17 km (11 miles/58,080 feet) at the equator. It is at its highest level over the equator and the lowest over the geographical north pole and south pole. On account of this, the coolest layer in the atmosphere lies at about 17 km over the equator. Due to the variation in starting height, the tropopause extremes are referred to as the equatorial tropopause and the polar tropopause.

Measuring the lapse rate through the troposphere and the stratosphere identifies the location of the tropopause. In the troposphere, the lapse rate is, on average, 6.5 °C per kilometre in the absence of inversions. In the stratosphere, however, the temperature increases with altitude. Alternatively, a dynamic definition of the tropopause is used with potential vorticity instead of vertical temperature gradient as the defining variable. There is no universally used threshold: the most common ones are: the tropopause lies at the 2 PVU or 1.5 PVU surface. PVU stands for *potential vorticity unit* (PVU). This threshold will be taken as a positive or negative value (e.g. 2 and "2 PVU), giving surfaces located in the northern and southern hemisphere respectively. To define a global tropopause in this way, the two surfaces arising from the positive and negative thresholds need to be joined near the equator using another type of surface such as a constant potential temperature surface.

It is also possible to define the tropopause in terms of chemical composition. For example, the lower stratosphere has much higher ozone concentrations than the upper troposphere, but much lower water vapour concentrations, so appropriate cutoffs can be used.

The tropopause is not a "hard" boundary. Such oscillation sets up a low-frequency atmospheric gravity wave capable of affecting both atmospheric and oceanic currents in the region.

Most commercial aircraft are flown below the tropopause or "trop" if at all possible to take advantage of the troposphere's temperature lapse rate. Jet engines are more efficient at lower temperatures.

Hydrosphere: The total mass of the Earth's hydrosphere is about 1.4×10^{18} tonnes, which is about 0.023% of the Earth's total mass. About 20×10^{12} tonnes of this is in the Earth's atmosphere (the volume of one tonne of water is approximately 1 cubic metre). Approximately 75% of the Earth's surface, an area of some 361 million square kilometres (139.5 million square miles), is covered by ocean. The average salinity of the Earth's oceans is about 35 grams of salt per kilogram of sea water (35%).

Other Hydrospheres: A thick hydrosphere is thought to exist around the Jovian moon Europa. The outer layer of this hydrosphere is almost entirely ice, but current models predict that there is an ocean up to 100 km in depth underneath the ice. This ocean remains in a liquid form because of tidal flexing of the moon in its orbit around Jupiter. The volume of Europa's hydrosphere is 3×10^{18} m^3, 2.3 times that of Earth.

It has been suggested that the Jovian moon Ganymede and the Saturnian moon Enceladus may also possess sub-surface oceans. However the ice covering is expected to be thicker on Jupiter's Ganymede than on Europa.

Hydrological Cycle: Insolation, or energy (in the form of heat and light) from the sun, provides the energy necessary to cause evaporation from all wet surfaces including oceans, rivers, lakes, soil and the leaves of plants. Water vapour is further released as transpiration from vegetation and from humans and other animals.

Cryosphere: The cryosphere is the term which collectively describes the portions of the Earth's surface where water is in solid form, including sea ice, lake ice, river ice, snow cover, glaciers, ice caps and ice sheets, and frozen ground (which includes permafrost). Thus there is a wide overlap with the hydrosphere. The cryosphere is an integral part of the global climate system with important linkages and feedbacks generated through its influence on surface energy and moisture fluxes, clouds, precipitation, hydrology, atmospheric and oceanic circulation. Through these feedback processes, the cryosphere plays a significant role in global climate and in climate model response to global change.

Structure: Frozen water is found on the Earth's surface primarily as snow cover, freshwater ice in lakes and rivers, sea ice, glaciers, ice sheets, and frozen ground and permafrost (permanently-frozen ground). The residence time of water in each of these cryospheric sub-systems varies widely. Snow cover and freshwater ice are essentially seasonal, and most sea ice, except for ice in the central Arctic, lasts only a few years if it is not seasonal. A given water particle in glaciers, ice sheets, or ground ice, however, may remain frozen for 10-100,000 years or longer, and deep ice in parts of East Antarctica may have an age approaching 1 million years. Most of the world's ice volume is in Antarctica, principally in the East Antarctic Ice Sheet. In terms of areal extent, however, Northern Hemisphere winter snow and ice extent comprise the largest area, amounting to an average 23% of hemispheric surface area in January. The large areal extent and the important climatic roles of snow and ice, related to their unique physical properties, indicate that the ability to observe and model snow and ice-cover extent, thickness, and physical properties (radiative and thermal properties) is of particular significance for climate research.

There are several fundamental physical properties of snow and ice that modulate energy exchanges between the surface and the atmosphere. The most important properties are the surface reflectance (albedo), the ability to transfer heat (thermal diffusivity), and the ability to change state (latent heat). These physical properties, together with surface roughness, emissivity, and dielectric characteristics, have important implications for observing snow and ice from space. For example, surface roughness is often the dominant factor determining the strength of radar backscatter. Physical properties such as crystal structure, density, length, and liquid-water content are important factors affecting the transfers of heat and water and the scattering of microwave energy.

The surface reflectance of incoming solar radiation is important for the surface energy balance (SEB). It is the ratio of reflected to incident solar radiation, commonly referred to as albedo. Climatologists are primarily interested in albedo integrated over the shortwave portion of the electromagnetic spectrum (~0.3 to 3.5 nm), which coincides with the main solar energy input. Typically, albedo values for non-melting snow-covered surfaces are high (~80-90%) except in the case of forests.

The higher albedos for snow and ice cause rapid shifts in surface reflectivity in autumn and spring in high latitudes, but the overall climatic significance of this increase is spatially and temporally modulated by cloud cover. (Planetary albedo is determined principally by cloud cover, and by the small amount of total solar radiation received in high latitudes during winter months.) Summer and autumn are times of high-average cloudiness over the Arctic Ocean so the albedo feedback associated with the large seasonal changes in sea-ice extent is greatly reduced. Groisman *et al.* (1994a) observed that snow cover exhibited the greatest influence on the Earth radiative balance in the spring (April to May) period when incoming solar radiation was greatest over snow-covered areas.

The thermal properties of cryospheric elements also have important climatic consequences. Snow and ice have much lower thermal diffusivities than air. Thermal diffusivity is a measure of the speed at which temperature waves can penetrate a substance. Snow and ice are many orders of magnitude less efficient at diffusing heat than air. Snow cover insulates the ground surface, and sea ice insulates the underlying ocean, decoupling the surface-atmosphere interface with respect to both heat and moisture fluxes.

The flux of moisture from a water surface is eliminated by even a thin skin of ice, whereas the flux of heat through thin ice continues to be substantial until it attains a thickness in excess of 30 to 40 cm. However, even a small amount of snow on top of the ice will dramatically reduce the heat flux and slow down the rate of ice growth. The insulating effect of snow also has major implications for the hydrological cycle. In non-permafrost regions, the insulating effect of snow is such that only near-surface ground freezes and deep-water drainage is uninterrupted.

While snow and ice act to insulate the surface from large energy losses in winter, they also act to retard warming in the spring and summer because of the large amount of energy required to melt ice (the latent heat of fusion, 3.34×10^5 J/kg at 0°C). However, the strong static stability of the atmosphere over areas of extensive snow or ice tends to confine the immediate cooling effect to a relatively shallow layer, so that associated atmospheric anomalies are usually short-lived and local to regional in scale. In some areas of the world such as Eurasia, however, the cooling associated with a heavy snowpack and moist spring soils is known to play a role in modulating the summer monsoon circulation. Gutzler and Preston (1997) recently presented evidence for a similar snow-summer circulation feedback over the southwestern United States.

The role of snow cover in modulating the monsoon is just one example of a short-term cryosphere-climate feedback involving the land surface and the atmosphere. These operate over a wide range of spatial and temporal scales from local seasonal cooling of air temperatures to hemispheric-scale variations in ice sheets over time-scales of thousands of years. The feedback mechanisms involved are often complex and incompletely understood. For example, Curry *et al.* (1995) showed that the so-called "simple" sea ice-albedo feedback involved complex interactions with lead fraction, melt ponds, ice thickness, snow cover, and sea-ice extent.

Snow: Snow cover has the second-largest areal extent of any component of the cryosphere, with a mean maximum areal extent of approximately 47 million km^2. Most of the Earth's snow-covered area (SCA) is located in the Northern Hemisphere, and temporal variability is dominated by the seasonal cycle; Northern Hemisphere snow-cover extent ranges from 46.5 million km^2 in January to 3.8 million km^2 in August. North American winter SCA has exhibited an increasing trend over much of this century largely in response to an increase in precipitation.

However, the available satellite data show that the hemispheric winter snow cover has exhibited little interannual variability over the 1972-1996 period, with a coefficient of variation (COV=s.d./mean) for January Northern Hemisphere snow cover of < 0.04. According to Groisman *et al.* (1994a) Northern Hemisphere spring snow cover should exhibit a decreasing trend to explain an observed increase in Northern Hemisphere spring air temperatures this century. Preliminary estimates of SCA from historical and reconstructed in situ snow-cover data suggest this is the case for Eurasia, but not for North America, where spring snow cover has remained close to current levels over most of this century. Because of the close relationship observed between hemispheric air temperature and snow-cover extent over the period of satellite data (IPCC 1996), there is considerable interest in monitoring Northern Hemisphere snow-cover extent for detecting and monitoring climate change.

Snow cover is an extremely important storage component in the water balance, especially seasonal snowpacks in mountainous areas of the world. Though limited in extent, seasonal snowpacks in the Earth's mountain ranges account for the major source of the runoff for stream flow and groundwater recharge over wide areas of the midlatitudes. For example, over 85% of the annual runoff from the Colorado River basin originates as snowmelt. Snowmelt runoff from the Earth's mountains fills the rivers and recharges the aquifers that over a billion people depend on for their water resources. Further, over 40% of the world's protected areas are in mountains, attesting to their value both as unique ecosystems needing protection and as recreation areas for humans. Climate warming is expected to result in major changes to the partitioning of snow and rainfall, and to the timing of snowmelt, which will have important implications for water use and management.

These changes also involve potentially important decadal and longer time-scale feedbacks to the climate system through temporal and spatial changes in soil moisture and runoff to the oceans.(Walsh 1995). Freshwater fluxes from the snow cover into the marine environment may be important, as the total flux is probably of the same magnitude as desalinated ridging and rubble areas of sea ice. In addition, there is an associated pulse of precipitated pollutants which accumulate over the Arctic winter in snowfall and are released into the ocean upon ablation of the sea-ice.

Sea Ice: Sea ice covers much of the polar oceans and forms by freezing of sea water. Satellite data since the early 1970s reveal considerable seasonal, regional, and interannual variability in the sea-ice covers of both hemispheres. Seasonally, sea-ice extent in the Southern Hemisphere varies by a factor of 5, from a minimum of 3-4 million km^2 in February to a maximum of 17-20 million km^2 in September. The seasonal variation is much less in the Northern Hemisphere where the confined nature and high latitudes of the Arctic Ocean result in a much larger perennial ice cover, and the surrounding land limits the equatorward extent of wintertime ice. Thus, the seasonal variability in Northern Hemisphere ice extent varies by only a factor of 2, from a minimum of 7-9 million km^2 in September to a maximum of 14-16 million km^2 in March.

The ice cover exhibits much greater regional-scale interannual variability than it does hemispherical. For instance, in the region of the Seas of Okhotsk and Japan, maximum ice extent decreased from 1.3 million km^2 in 1983 to 0.85 million km^2 in 1984, a decrease of 35%, before rebounding the following year to 1.2 million km^2. The regional fluctuations in both hemispheres are such that for any several-year period of the satellite record some regions exhibit decreasing ice coverage while others exhibit increasing ice cover. The overall trend indicated in the passive microwave record from 1978 through mid-1995 shows that the extent of Arctic sea ice is decreasing 2.7% per decade. Subsequent work with the satellite passive-microwave data indicates that from late October 1978 through the end of 1996 the extent of Arctic sea ice decreased by 2.9% per decade while the extent of Antarctic sea ice increased by 1.3% per decade.

Lake Ice and River Ice: Ice forms on rivers and lakes in response to seasonal cooling. The sizes of the ice bodies involved are too small to exert other than localized climatic effects. However, the freeze-up/ break-up processes respond to large-scale and local weather factors, such that considerable interannual variability exists in the dates of appearance and disappearance of the ice. Long series of lake-ice observations can serve as a proxy climate record, and the monitoring of freeze-up and break-up trends may provide a convenient integrated and seasonally specific index of climatic perturbations. Information on river-ice conditions is less useful as a climatic proxy because ice formation is strongly dependent on river-flow regime, which is affected by precipitation, snow melt, and watershed runoff as well as being subject to human interference that directly modifies channel flow, or that indirectly affects the runoff via land-use practices.

Lake freeze-up depends on the heat storage in the lake and therefore on its depth, the rate and temperature of any inflow, and water-air energy fluxes. Information on lake depth is often unavailable, although some indication of the depth of shallow lakes in the Arctic can be obtained from airborne radar imagery during late winter (Sellman *et al.* 1975) and spaceborne optical imagery during summer. The timing of breakup is modified by snow depth on the ice as well as by ice thickness and freshwater inflow.

Frozen Ground and Permafrost: Frozen ground (permafrost and seasonally frozen ground) occupies approximately 54 million km^2 of the exposed land areas of the Northern Hemisphere and therefore has the largest areal extent of any component of the cryosphere. Permafrost (perennially frozen ground) may occur where mean annual air temperatures (MAAT) are less than -1 or -2°C and is generally continuous where MAAT are less than -7°C. In addition, its extent and thickness are affected by ground moisture content, vegetation cover, winter snow depth, and aspect. The global extent of permafrost is still not completely known, but it underlies approximately 20% of Northern Hemisphere land areas. Thicknesses exceed 600 m along the Arctic coast of northeastern Siberia and Alaska, but, toward the margins, permafrost becomes thinner and horizontally discontinuous. The marginal zones will be more immediately subject to any melting caused by a warming trend. Most of the presently existing permafrost formed during previous colder conditions and is therefore relic. However, permafrost may form under present-day polar climates where glaciers retreat or land emergence exposes unfrozen ground. Washburn (1973) concluded that most continuous permafrost is in balance with the present climate at its upper surface, but changes at the base depend on the present climate and geothermal heat flow; in contrast, most discontinuous permafrost is probably unstable or "in such delicate equilibrium that the slightest climatic or surface change will have drastic disequilibrium effects".

Under warming conditions, the increasing depth of the summer active layer has significant impacts on the hydrologic and geomorphic regimes. Thawing and retreat of permafrost have been reported in the upper Mackenzie Valley and along the southern margin of its occurrence in Manitoba, but such observations are not readily quantified and generalized. Based on average latitudinal gradients of air temperature, an average northward displacement of the southern permafrost boundary by 50-to-150 km could be expected, under equilibrium conditions, for a 1°C warming.

Only a fraction of the permafrost zone consists of actual ground ice. The remainder (dry permafrost) is simply soil or rock at subfreezing temperatures. The ice volume is generally greatest in the uppermost permafrost layers and mainly comprises pore and segregated ice in Earth material. Measurements of bore-hole temperatures in permafrost can be used as indicators of net changes in temperature regime. Gold and Lachenbruch (1973) infer a 2-4°C warming over 75 to 100 years at Cape Thompson, Alaska, where the upper 25% of the 400-m thick permafrost is unstable with respect to an equilibrium profile of temperature with depth (for the present mean annual surface temperature of -5°C). Maritime influences may have biased this estimate, however. At Prudhoe Bay similar data imply a 1.8°C warming over the last 100 years. Further complications may be introduced by changes in snow-cover depths and the natural or artificial disturbance of the surface vegetation.

The potential rates of permafrost thawing have been established by Osterkamp (1984) to be two centuries or less for 25-meter-thick permafrost in the discontinuous zone of interior Alaska, assuming warming from -0.4 to 0°C in 3–4 years, followed by a further 2.6°C rise. Although the response of permafrost (depth) to temperature change is typically a very slow process, there is ample evidence for the fact that the active layer thickness quickly responds to a temperature change. Whether, under a warming or cooling scenario, global climate change will have a significant effect on the duration of frost-free periods in both regions with seasonally-and perennially-frozen ground.

Glaciers and ice Sheets: Ice sheets and glaciers are flowing ice masses that rest on solid land. They are controlled by snow accumulation, surface and basal melt, calving into surrounding oceans or lakes and internal dynamics. The latter results from gravity-driven creep flow ("glacial flow") within the ice body and sliding on the underlying land, which leads to thinning and horizontal spreading. Any imbalance of this dynamic equilibrium between mass gain, loss and transport due to flow results in either growing or shrinking ice bodies.

Ice sheets are the greatest potential source of global freshwater, holding approximately 77% of the global total. This corresponds to 80 m of world sea-level equivalent, with Antarctica accounting for 90% of this. Greenland accounts for most of the remaining 10%, with other ice bodies and glaciers accounting for less than 0.5%. Because of their size in relation to annual rates of snow accumulation and melt, the residence time of water in ice sheets can extend to 100,000 or 1 million years. Consequently, any climatic perturbations produce slow responses, occurring over glacial and interglacial periods. Valley glaciers respond rapidly to climatic fluctuations with typical response times of 10–50 years. However, the response of individual glaciers may be asynchronous to the same climatic forcing because of differences in glacier length, elevation, slope, and speed of motion. Oerlemans (1994) provided evidence of coherent global glacier retreat which could be explained by a linear warming trend of 0.66°C per 100 years.

While glacier variations are likely to have minimal effects upon global climate, their recession may have contributed one third to one half of the observed 20th Century rise in sea level. Furthermore, it is extremely likely that such extensive glacier recession as is currently observed in the Western Cordillera of North America, where runoff from glacierized basins is used for irrigation and hydropower, involves significant hydrological and ecosystem impacts. Effective water-resource planning and impact mitigation in such areas depends upon developing a sophisticated knowledge of the status of glacier ice and the mechanisms that cause it to change. Furthermore, a clear understanding of the mechanisms at work is crucial to interpreting the global-change signals that are contained in the time series of glacier mass balance records.

Combined glacier mass balance estimates of the large ice sheets carry an uncertainty of about 20%. Studies based on estimated snowfall and mass output tend to indicate that the ice sheets are near balance or taking some water out of the oceans. Marinebased studies suggest sea-level rise from the Antarctic or rapid ice-shelf basal melting. Some authors have suggested that the difference between the observed rate of sea-level rise (roughly 2 mm/y) and the explained rate of sea-level rise from melting of mountain glaciers, thermal expansion of the ocean, etc. (roughly 1 mm/y or less) is similar to the modelled imbalance in the Antarctic, suggesting a contribution of sea-level rise from the Antarctic.

Relationships between global climate and changes in ice extent are complex. The mass balance of land-based glaciers and ice sheets is determined by the accumulation of snow, mostly in winter, and warm-season ablation due primarily to net radiation and turbulent heat fluxes to melting ice and snow from warm-air advection,(Munro 1990). However, most of Antarctica never experiences surface melting. Where ice masses terminate in the ocean, iceberg calving is the major contributor to mass loss. In this situation, the ice margin may extend out into deep water as a floating ice shelf, such as that in the Ross Sea. Despite the possibility that global warming could result in losses to the Greenland ice sheet being offset by gains to the Antarctic ice sheet, there is major concern about the possibility of a West Antarctic Ice Sheet collapse. The West Antarctic Ice Sheet is grounded on bedrock below sea level, and its collapse has the potential of raising the world sea level 6–7 m over a few hundred years. Most of the discharge of the West Antarctic Ice Sheet is via the five major ice streams (faster flowing ice) entering the Ross Ice Shelf, the Rutford Ice Stream entering Ronne-Filchner shelf of the Weddell Sea, and the Thwaites Glacier and Pine Island Glacier entering the Amundsen Ice Shelf.

Opinions differ as to the present mass balance of these systems, principally because of the limited data. The West Antarctic Ice Sheet is stable so long as the Ross Ice Shelf is constrained by drag along its lateral boundaries and pinned by local grounding.

Lithosphere: The lithosphere is the rigid outermost shell of a rocky planet. It comprises the crust and the portion of the upper mantle that behaves elastically on time scales of thousands of years or greater.

Earth's Lithosphere: In the Earth, the lithosphere includes the crust and the uppermost mantle, which constitute the hard and rigid outer layer of the Earth. The lithosphere is underlain by the asthenosphere, the weaker, hotter, and deeper part of the upper mantle. The boundary between the lithosphere and the underlying asthenosphere is defined by a difference in response to stress: the lithosphere remains rigid for very long periods of geologic time in which it deforms elastically and through brittle failure, while the asthenosphere deforms viscously and accommodates strain through plastic deformation. The lithosphere is broken into tectonic plates. The uppermost part of the lithosphere that chemically reacts to the atmosphere, hydrosphere and biosphere through the soil forming process is called the pedosphere.

The concept of the lithosphere as Earth's strong outer layer was developed by Joseph Barrell, who wrote a series of papers introducing the concept. The concept was based on the presence of significant gravity anomalies over continental crust, from which he inferred that there must exist a strong upper layer (which he called the lithosphere) above a weaker layer which could flow (which he called the asthenosphere). These ideas were expanded by Daly (1940), and have been broadly accepted by geologists and geophysicists. Although these ideas about lithosphere and asthenosphere were developed long before plate tectonic theory was articulated in the 1960s, the concepts that a strong lithosphere exists and that this rests on a weak asthenosphere are essential to that theory.

The lithosphere provides a conductive lid atop the convecting mantle; as such, it affects heat transport through the Earth.

There are two types of lithosphere:

- Oceanic lithosphere, which is associated with Oceanic crust and exists in the ocean basins
- Continental lithosphere, which is associated with Continental crust.

The thickness of the lithosphere is considered to be the depth to the isotherm associated with the transition between brittle and viscous behaviour. The temperature at which olivine begins to deform viscously (~1000°C) is often used to set this isotherm because olivine is generally the weakest mineral in the upper mantle. Oceanic lithosphere is typically about 50–100 km thick (but beneath the mid-ocean ridges is no thicker than the crust), while continental lithosphere has a range in thickness from about 40 km to perhaps 200 km; the upper ~30 to ~50 km of typical continental lithosphere is crust. The mantle part of the lithosphere consists largely of peridotite. The crust is distinguished from the upper mantle by the change in chemical composition that takes place at the Moho discontinuity.

Oceanic Lithosphere: Oceanic lithosphere consists mainly of mafic crust and ultramafic mantle (peridotite) and is denser than continental lithosphere, for which the mantle is associated with crust made of felsic rocks. Oceanic lithosphere thickens as it ages and moves away from the mid-ocean ridge. This thickening occurs by conductive cooling, which converts hot asthenosphere into lithospheric mantle, and causes the oceanic lithosphere to become increasingly thick and dense with age. The thickness of the mantle part of the oceanic lithosphere can be approximated as a thermal boundary layer that thickens as the square root of time.

Here, h is the thickness of the oceanic mantle lithosphere, κ is the thermal diffusivity (approximately 10^6 m^2/s), and t is time.

Oceanic lithosphere is less dense than asthenosphere for a few tens of millions of years, but after this becomes increasingly denser than asthenosphere. This is because the chemically-differentiated oceanic crust is lighter than asthenosphere, but due to thermal contraction, the mantle lithosphere is more dense than the asthenosphere. The gravitational instability of mature oceanic lithosphere has the effect that at subduction zones, oceanic lithosphere invariably sinks underneath the overriding lithosphere, which can be oceanic or continental. New oceanic lithosphere is constantly being produced at mid-ocean ridges and is recycled back to the mantle at subduction zones. As a result, oceanic lithosphere is much younger than continental lithosphere: the oldest oceanic lithosphere is about 170 million years old, while parts of the continental lithosphere are billions of years old. The oldest parts of continental lithosphere underlie cratons, and the mantle lithosphere there is thicker and less dense than typical; the relatively low density of such mantle "roots of cratons" helps to stabilize these regions.

Subducted Lithosphere: Geophysical studies in the early 21st Century posit that large pieces of the lithosphere have been subducted into the mantle as deep as 2900 km to near the core-mantle boundary, while others "float" in the upper mantle, while some stick down into the mantle as far as 400 km but remain "attached" to the continental plate above, similar to the extent of the "tectosphere" proposed by Jordan in 1988.

Mantle Xenoliths: Geoscientists can directly study the nature of the subcontinental mantle by examining mantle xenoliths brought up in kimberlite, lamproite, and other volcanic pipes. The histories of these xenoliths have been investigated by many methods, including analyses of abundances of isotopes of osmium and rhenium. Such studies have confirmed that mantle lithospheres below some cratons have persisted for periods in excess of 3 billion years, despite the mantle flow that accompanies plate tectonics.

Pedosphere: The pedosphere is the outermost layer of the Earth that is composed of soil and subject to soil formation processes. It exists at the interface of the lithosphere, atmosphere, hydrosphere and biosphere.

The pedosphere acts as the mediator of chemical and biogeochemical flux into and out of these respective systems and is made up of gaseous, mineralic, fluid and biologic components. The pedosphere lies within the Critical Zone, a broader interface that includes vegetation, pedosphere, groundwater aquifer systems, regolith and finally ends at some depth in the bedrock where the biosphere and hydrosphere cease to make significant changes to the chemistry at depth. As part of the larger global system, any particular environment in which soil forms is influenced solely by its geographic position on the globe as climatic, geologic, biologic and anthropogenic changes occur with changes in longitude and latitude.

The pedosphere lies below the vegetative cover of the biosphere and above the hydrosphere and lithosphere. The soil forming process (pedogenesis) can begin without the aid of biology but is significantly quickened in the presence of biologic reactions. Soil formation begins with the chemical and/or physical breakdown of minerals to form the initial material that overlies the bedrock substrate. Biology quickens this by secreting acidic compounds (dominantly fulvic acids) that help break rock apart. Particular biologic pioneers are lichen, mosses and seed bearing plants but many other inorganic reactions take place that diversify the chemical makeup of the early soil layer. Once weathering and decomposition products accumulate, a coherent soil body allows the migration of fluids both vertically and laterally through the soil profile causing ion exchange between solid, fluid and gaseous phases. As time progresses, the bulk geochemistry of the soil layer will deviate away from the initial composition of the bedrock and will evolve to a chemistry that reflects the type of reactions that take place in the soil.

Lithosphere: The primary conditions for soil development are controlled by the chemical composition of the rock that the soil will eventually be forming on. Rock types that form the base of the soil profile are often either sedimentary (carbonate or siliceous), igneous or metaigneous (metamorphosed igneous rocks) or volcanic and metavolcanic rocks. The rock type and the processes that lead to its exposure at the surface are controlled by the regional geologic setting of the specific area under study, which revolve around the underlying theory of plate tectonics, subsequent deformation, uplift, subsidence and deposition. Metaigneous and metavolcanic rocks form the largest component of cratons and are high in silica. Igneous and volcanic rocks are also high in silica but with non-metamorphosed rock, weathering becomes faster and the mobilization of ions is more widespread. Rocks high in silica produce silicic acid as a weathering product. There are few rock types that lead to localized enrichment of some of the biologically limiting elements like phosphorus (P) and nitrogen (N). Phosphatic shale ($<15\%$ P_2O_5) and phosphorite ($>15\%$ P_2O_5) form in anoxic deep water basins that preserve organic material. Greenstone (metabasalt), phyllite and schist release up to 30-50% of the nitrogen pool. Thick successions of carbonate rocks are often deposited on craton margins during sea level rise. The widespread dissolution of carbonate and evaporate minerals leads to elevated levels of Mg^{2+}, HCO_3^-, Sr^{2+}, Na^+, Cl^- and SO_4^{2-} ions in aqueous solution.

Weathering and Dissolution of Minerals: The process of soil formation is dominated by chemical weathering of silicate minerals, aided by acidic products of pioneering plants and organisms as well as carbonic acid inputs from the atmosphere. Carbonic acid is produced in the atmosphere and soil layers through the carbonation reaction.

$$H_2O + CO_2 \rightarrow H^+ + HCO_3^- \rightarrow H_2CO_3$$

This is the dominant form of chemical weathering and aides in the breakdown of carbonate minerals like calcite and dolomite and silicate minerals like feldspar. The breakdown of the Na-feldspar, albite, by carbonic acid to form kaolinite clay is as follows:

$$2NaAlSi_3O_8 + 2H_2CO_3 + 9H_2O \rightarrow 2Na^+ + 2HCO_3^- + 4H_4SiO_4 + Al_2Si_2O_5(OH)_4$$

Evidence of this reaction in the field would be elevated levels of bicarbonate (HCO_3^-), sodium and silica ions in the water runoff. The breakdown of carbonate minerals:

$$CaCO_3 + H_2CO_3 \rightarrow Ca^{2+} + 2HCO_3 \text{ or } CaCO_3 \rightarrow Ca^{2+} + CO_3^{2-}$$

The further dissolution of carbonic acid (H_2CO_3) and bicarbonate (HCO_3) produces CO_2 gas. Oxidization is also a major contributor to the breakdown of many silicate minerals and formation of secondary minerals (diagenesis) in the early soil profile. Oxidation of Olivine ($FeMgSiO_2$) releases Fe, Mg and Si ions. The Mg is soluble in water and is carried in the runoff but the Fe often reacts with oxygen to precipitate Fe_2O_3 (hematite), the oxidized state of iron oxide. Sulfur, a byproduct of decaying organic material will also react to Fe to form pyrite (FeS_2) but often in reducing environments. Pyrite dissolution leads to high pH levels due to elevated H+ ions and further precipitation of Fe_2O_3 ultimately changing the redox conditions of the environment.

Biosphere: Inputs from the biosphere may begin with lichen and other microorganisms that secrete oxalic acid. These microorganisms, associated with the lichen community or independently inhabiting rocks, include a number of blue-green algae, green algae, various fungi, and numerous bacteria. Lichen has long been viewed as the pioneers of soil development as the following statement suggests:

"The initial conversion of rock into soil is carried on by the pioneer lichens and their successors, the mosses, in which the hair-like rhizoids assume the role of roots in breaking down the surface into fine dust"

However, lichens are not necessarily the only pioneering organisms nor the earliest form of soil formation as it has been documented that seed-bearing plants may occupy an area and colonize quicker than lichen. Also, eolian sedimentation can produce high rates of sediment accumulation. Nonetheless, lichen can certainly withstand harsher conditions than most vascular plants and although they have slower colonization rates, do form the dominant group in alpine regions.

Acids released from plant roots include acetic and citric acids. During the decay of organic matter Phenolic acids are released from plant matter and humic and fulvic acids are released by soil microbes. These organic acids speed up chemical weathering by combining with some of the weathering products in a process known as chelation. In the soil profile, the organic acids are often concentrated at the top while carbonic acid plays a larger role towards the bottom or below in the aquifer.

As the soil column develops further into thicker accumulations, larger animals come to inhabit the soil and continue to alter the chemical evolution of their respective niche. Earthworms aerate the soil and convert large amounts of organic matter into rich humus, improving soil fertility. Small burrowing mammals store food, grow young and may hibernate in the pedosphere altering the course of soil evolution. Large mammalian herbivores above ground transport nutrients in form of nitrogen-rich waste and phosphorus-rich antlers while predators leave phosphorus-rich piles of bones on the soil surface, leading the localized enrichment of the soil below.

Redox Conditions in Wetland Soils: Nutrient cycling in lakes and freshwater wetlands depends heavily on redox conditions. Under a few millimeters of water heterotrophic bacteria metabolize and consume oxygen. They therefore deplete the soil of oxygen and create the need for anaerobic respiration.

Some anaerobic microbial processes include denitrification, sulfate reduction and methanogenesis and are responsible for the release of N_2 (nitrogen), H_2S (hydrogen sulfide) and CH_4 (methane). Other anaerobic microbial processes are linked to changes in the oxidation state of iron and manganese. As a result of anaerobic decomposition, the soil stores large amounts of organic carbon because decomposition is incomplete. The redox potential describes which way chemical reactions will proceed in oxygen deficient soils and controls the nutrient cycling in flooded systems. Redox potential, or reduction potential, is used to express the likelihood of an environment to receive electrons and therefore become reduced. For example, if a system already has plenty of electrons (anoxic, organic-rich shale) it is reduced and will likely donate electrons to a part of the system that has a low concentration of electrons, or an oxidized environment, to equilibrate to the chemical gradient. The oxidized environment has high redox potential, whereas the reduced environment has a low redox potential.

The redox potential is controlled by the oxidation state of the chemical species, pH and the amount of oxygen (O_2) there is in the system. The oxidizing environment accepts electrons because of the presence of O_2, which acts as electron acceptors:

$$O_2 + 4e^- + 4H^+ \rightarrow H_2O$$

This equation will tend to move to the right in acidic conditions which causes higher redox potentials to be found at lower pH levels. Bacteria, heterotrophic organisms, consume oxygen while decomposing organic material which depletes the soils of oxygen, thus increasing the redox potential. In low redox conditions the deposition of ferrous iron $(Fe)^{2+}$ will increase with decreasing decomposition rates, thus preserving organic remains and depositing humus.

At high redox potential, the oxidized form of iron, ferric iron (Fe^{3+}), will be deposited commonly as hematite. By using analytical geochemical tools such as x-ray fluorescence (XRF) or inductively coupled mass spectroscopy (ICP-MS) the two forms of Fe (Fe^{2+} and Fe^{3+}) can be measured in ancient rocks therefore determining the redox potential for ancient soils.

Such a study was done on Permian through Triassic rocks (300-200 million years old) in Japan and British Colombia. The geologists found hematite throughout the early and middle Permian but began to find the reduced form of iron in pyrite within the ancient soils near the end of the Permian and into the Triassic. This suggests that conditions became less oxygen rich, even anoxic, during the late Permian which eventually lead to the greatest extinction in earth's history, the P-T extinction. Decomposition in anoxic or reduced soils is also carried out by sulfur-reducing bacteria which, instead of O_2 use SO_4^{2-} as an electron acceptor and produce hydrogen sulfide (H_2S) and carbon dioxide in the process:

$$2H^+ + SO_4^{2-} + 2(CH_2O) \rightarrow 2CO_2 + H_2S + 2H_2O$$

The H_2S gas percolates upwards and reacts with Fe^{2+} and precipitates pyrite, acting as a trap for the toxic H_2S gas. However, H_2S is still a large fraction of emissions from wetland soils. In most freshwater wetlands there is little sulfate (SO_4^{2-}) so methanogenesis becomes the dominant form of decomposition by methanogenic bacteria only when sulfate is depleted. Acetate, a compound that is a byproduct of fermenting cellulose is split by methanogenic bacteria to produce methane (CH_4) and carbon dioxide (CO_2), which are released to the atmosphere. Methane is also released during the reduction of CO_2 by the same bacteria.

Atmosphere: In the pedosphere it is safe to assume that gases are in equilibrium with the atmosphere. Because plant roots and soil microbes release CO_2 to the soil, the concentration of bicarbonate(HCO_3) in soil waters is much greater than that in equilibrium with the atmosphere, the high concentration of CO_2 and the occurrence of metals in soil solutions results in lower pH levels in the soil. Gases that escape from the pedosphere to the atmosphere include the gaseous byproducts of carbonate dissolution, decomposition, redox reactions and microbial photosynthesis. The main inputs from the atmosphere are aeolian sedimentation, rainfall and gas diffusion. Eolian sedimentation includes anything that can be entrained by wind or that stays suspended, seemingly indefinitely, in air and includes a wide variety of aerosol particles, biological particles like pollen and dust to pure quartz sand. Nitrogen is the most abundant constituent in rain, as water vapour utilizes aerosol particles to nucleate rain droplets.

Soil in Forests: Soil is well developed in the forest as suggested by the thick humus layers, rich diversity of large trees and animals that live there. In forests, precipitation exceeds evapotranspiration which results in an excess of water that percolates downward through the soil layers. Slow rates of decomposition leads to large amounts of fulvic acid, greatly enhancing chemical weathering. The downward percolation, in conjunction with chemical weathering leaches magnesium (Mg), iron (Fe), and aluminum (Al) from the soil and transports them downward, a process known as podzolization. This process leads to marked contrasts in the appearance and chemistry of the soil layers.

Soil in Grasslands and Deserts: Precipitation in grasslands is equal to or less than evapotranspiration and causes soil development to operate in relative drought. Leaching and migration of weathering products is therefore decreased. Large amounts of evaporation causes buildup of calcium (Ca) and other large cations flocculate clay minerals and fulvic acids in the upper soil profile. Impermeable clay limits downward percolation of water and fulvic acids, reducing chemical weathering and podzolization.

The depth to the maximum concentration of clay increases in areas of increased precipitation and leaching. When leaching is decreased, the Ca precipitates as calcite ($CaCO_3$) in the lower soil levels, a layer known as caliche. Deserts behave similarly to grasslands but operate in constant drought as precipitation is less than evapotranspiration. Chemical weathering proceeds more slowly than in grasslands and beneath the caliche layer may be a layer of gypsum and halite.

To study soils in deserts, pedologists have used the concept of chronosequences to relate timing and development of the soil layers. It has been shown that P is leached very quickly from the system and therefore decreases with increasing age. Furthermore, carbon buildup in the soils is decrease due to slower decomposition rates. As a result, the rates of carbon circulation in the biogeochemical cycle is decreased.

Physical Properties of Atmosphare

Pressure and Thickness

The average atmospheric pressure at sea level is about 1 atmosphere (atm) = 101.3 kPa (kilopascals) = 14.7 psi (pounds per square inch) = 760 torr = 29.92 inches of mercury (symbol Hg). Total atmospheric mass is 5.1480×1018 kg (1.135×1019 lb), about 2.5% less than would be inferred from the average sea level pressure and the Earth's area of 51007.2 megahectares, this portion being displaced by the Earth's mountainous terrain. Atmospheric pressure is the total weight of the air above unit area at the point where the pressure is measured. Thus air pressure varies with location and weather.

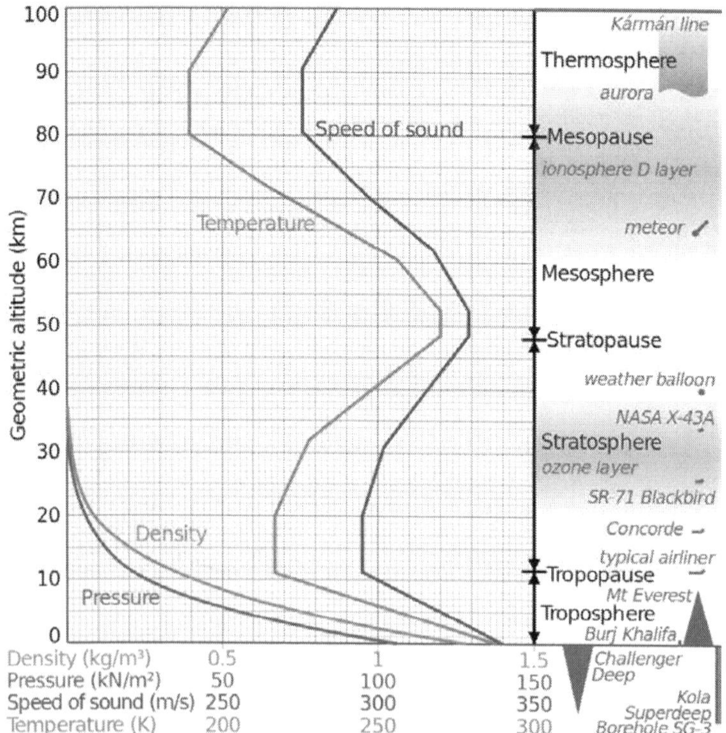

Figure: *Comparison of the 1962 US Standard Atmosphere graph of geometric altitude against air density, pressure, the speed of sound and temperature with approximate altitudes of various objects.*

If atmospheric density were to remain constant with height the atmosphere would terminate abruptly at 8.50 km (27,900 ft). Instead, density decreases with height, dropping by 50% at an altitude of about 5.6 km (18,000 ft). As a result the pressure decrease is approximately exponential with height, so that pressure decreases by a factor of two approximately every 5.6 km (18,000 ft) and by a factor of e = 2.718... approximately every 7.64 km (25,100 ft), the latter being the average scale height of Earth's atmosphere below 70 km (43 mi; 230,000 ft). However, because of changes in temperature, average molecular weight, and gravity throughout the atmospheric column, the dependence of atmospheric pressure on altitude is modeled by separate equations for each of the layers listed above. Even in the exosphere, the atmosphere is still present. This can be seen by the effects of atmospheric drag on satellites.

In summary, the equations of pressure by altitude in the above references can be used directly to estimate atmospheric thickness. However, the following published data are given for reference:

- 50% of the atmosphere by mass is below an altitude of 5.6 km (18,000 ft).
- 90% of the atmosphere by mass is below an altitude of 16 km (52,000 ft). The common altitude of commercial airliners is about 10 km (33,000 ft) and Mt. Everest's summit is 8,848 m (29,029 ft) above sea level.
- 99.99997% of the atmosphere by mass is below 100 km (62 mi; 330,000 ft), although in the rarefied region above this there are auroras and other atmospheric effects. The highest X-15 plane flight in 1963 reached an altitude of 108.0 km (354,300 ft).

Density and Mass

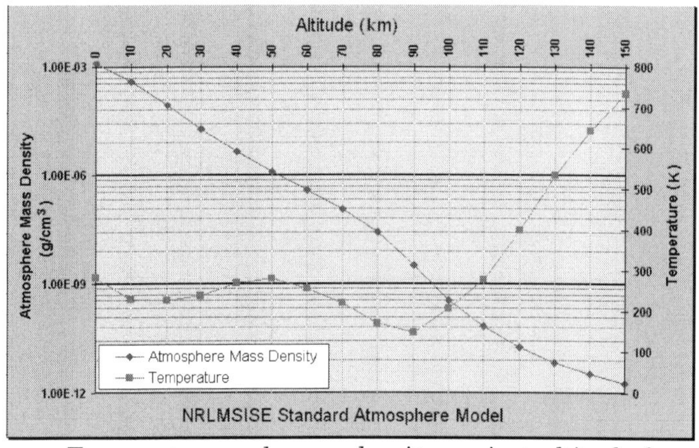

Figure: *Temperature and mass density against altitude from the NRLMSISE-00 standard atmosphere model (the eight dotted lines in each "decade" are at the eight cubes 8, 27, 64, ..., 729)*

The density of air at sea level is about 1.2 kg/m3 (1.2 g/L). Density is not measured directly but is calculated from measurements of temperature, pressure and humidity using the equation of state for air (a form of the ideal gas law). Atmospheric density decreases as the altitude increases. This variation can be approximately modeled using the barometric formula. More sophisticated models are used to predict orbital decay of satellites. The average mass of the atmosphere is about 5 quadrillion (5×1015) tonnes or 1/1,200,000 the mass of Earth. According to the American National Center for Atmospheric Research, "The total mean mass of the atmosphere is 5.1480×1018 kg with an annual range due to water vapor of 1.2 or 1.5×1015 kg depending on whether surface pressure or water vapor data are used; somewhat smaller than the previous estimate. The mean mass of water vapor is estimated as 1.27×1016 kg and the dry air mass as 5.1352 ±0.0003×1018 kg."

Optical Properties

Solar radiation (or sunlight) is the energy the Earth receives from the Sun. The Earth also emits radiation back into space, but at longer wavelengths that we cannot see. Part of the incoming and emitted radiation is absorbed or reflected by the atmosphere.

Scattering

When light passes through our atmosphere, photons interact with it through scattering. If the light does not interact with the atmosphere, it is called direct radiation and is what you see if you were to look directly at the Sun. Indirect radiation is light that has been scattered in the atmosphere. For example, on an overcast day when you cannot see your shadow there is no direct radiation reaching you, it has all been scattered. As another example, due to a phenomenon called Rayleigh scattering, shorter (blue) wavelengths scatter more easily than longer (red) wavelengths. This is why the sky looks blue, you are seeing scattered blue light. This is also why sunsets are red. Because the Sun is close to the horizon, the Sun's rays pass through more atmosphere than normal to reach your eye. Much of the blue light has been scattered out, leaving the red light in a sunset.

Absorption

Different molecules absorb different wavelengths of radiation. For example, O_2 and O3 absorb almost all wavelengths shorter than 300 nanometers. Water (H_2O) absorbs many wavelengths above 700 nm. When a molecule absorbs a photon, it increases the energy of the molecule. We can think of this as heating the atmosphere, but the atmosphere also cools by emitting radiation, as discussed below.

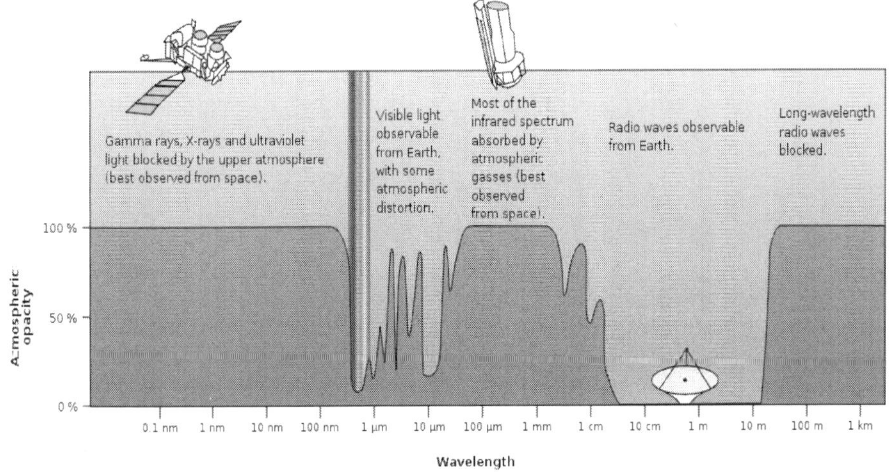

Figure: *Rough plot of Earth's atmospheric transmittance (or opacity) to various wavelengths of electromagnetic radiation, including visible light.*

The combined absorption spectra of the gases in the atmosphere leave "windows" of low opacity, allowing the transmission of only certain bands of light. The optical window runs from around 300 nm (ultraviolet-C) up into the range humans can see, the visible spectrum (commonly called light), at roughly 400–700 nm and continues to the infrared to around 1100 nm. There are also infrared and radio windows that transmit some infrared and radio waves at longer wavelengths. For example, the radio window runs from about one centimeter to about eleven-meter waves.

Emission

Emission is the opposite of absorption, it is when an object emits radiation. Objects tend to emit amounts and wavelengths of radiation depending on their "black body" emission curves, therefore hotter objects tend to emit more radiation, with shorter wavelengths. Colder objects emit less radiation, with longer wavelengths. For example, the Sun is approximately 6,000 K (5,730 °C; 10,340 °F), its radiation peaks near 500 nm, and is visible to the human eye. The Earth is approximately 290 K (17 °C; 62 °F), so its radiation peaks near 10,000 nm, and is much too long to be visible to humans.

Because of its temperature, the atmosphere emits infrared radiation. For example, on clear nights the Earth's surface cools down faster than on cloudy nights. This is because clouds (H_2O) are strong absorbers and emitters of infrared radiation. This is also why it becomes colder at night at higher elevations. The atmosphere acts as a "blanket" to limit the amount of radiation the Earth loses into space. The greenhouse effect is directly related to this absorption and emission (or "blanket") effect. Some chemicals in the atmosphere absorb and emit infrared radiation, but do not interact with sunlight in the visible spectrum. Common examples of these chemicals are CO_2 and H_2O. If there are too much of these greenhouse gases, sunlight heats the Earth's surface, but the gases block the infrared radiation from exiting back to space. This imbalance causes the Earth to warm, and thus climate change.

Refractive index

The refractive index of air is close to, but just greater than 1. Systematic variations in refractive index can lead to the bending of light rays over long optical paths. One example is that, under some circumstances, observers onboard ships can see other vessels just over the horizon because light is refracted in the same direction as the curvature of the Earth's surface.

The refractive index of air depends on temperature, giving rise to refraction effects when the temperature gradient is large. An example of such effects is the mirage.

Circulation

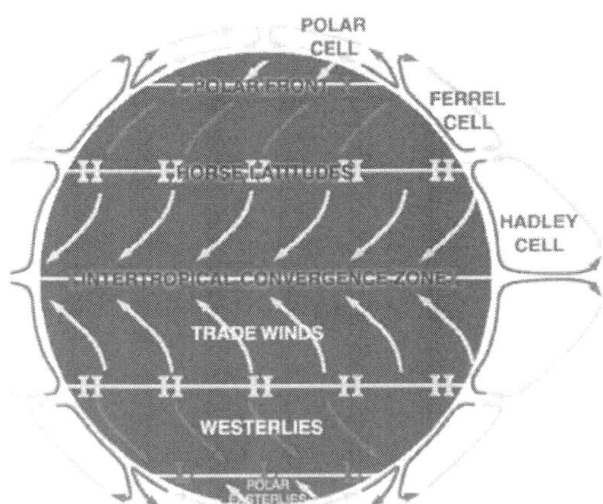

Figure: *An idealised view of three large circulation cells.*

Atmospheric circulation is the large-scale movement of air through the troposphere, and the means (with ocean circulation) by which heat is distributed around the Earth. The large-scale structure of the atmospheric circulation varies from year to year, but the basic structure remains fairly constant as it is determined by the Earth's rotation rate and the difference in solar radiation between the equator and poles.

Evolution of Earth's Atmosphere

Earliest Atmosphere

The outgassings of the Earth were stripped away by solar winds early in the history of the planet until a steady state was established, the first atmosphere. Based on today's volcanic evidence, this atmosphere would have contained 60% hydrogen, 20% oxygen (mostly in the form of water vapor), 10% carbon dioxide, 5 to 7% hydrogen sulfide, and smaller amounts of nitrogen, carbon monoxide, free hydrogen, methane and inert gases.

A major rainfall led to the buildup of a vast ocean, enriching the other agents, first carbon dioxide and later nitrogen and inert gases. A major part of carbon dioxide exhalations were soon dissolved in water and built up carbonate sediments.

Second Atmosphere

Water-related sediments have been found dating from as early as 3.8 billion years ago. About 3.4 billion years ago, nitrogen was the major part of the then stable "second atmosphere". An influence of life has to be taken into account rather soon in the history of the atmosphere, since hints of early life forms are to be found as early as 3.5 billion years ago. The fact that this is not perfectly in line with the 30% lower solar radiance (compared to today) of the early Sun has been described as the "faint young Sun paradox".

The geological record however shows a continually relatively warm surface during the complete early temperature record of the Earth with the exception of one cold glacial phase about 2.4 billion years ago. In the late Archaean eon an oxygen-containing atmosphere began to develop, apparently from photosynthesizing algae which have been found as stromatolite fossils from 2.7 billion years ago. The early basic carbon isotopy (isotope ratio proportions) is very much in line with what is found today, suggesting that the fundamental features of the carbon cycle were established as early as 4 billion years ago.

Third Atmosphere

The accretion of continents about 3.5 billion years ago added plate tectonics, constantly rearranging the continents and also shaping long-term climate evolution by allowing the transfer of carbon dioxide to large land-based carbonate storages. Free oxygen did not exist until about 1.7 billion years ago and this can be seen with the development of the red beds and the end of the banded iron formations. This signifies a shift from a reducing atmosphere to an oxidising atmosphere. O2 showed major ups and downs until reaching a steady state of more than 15%. The following time span was the Phanerozoic eon, during which oxygen-breathing metazoan life forms began to appear.

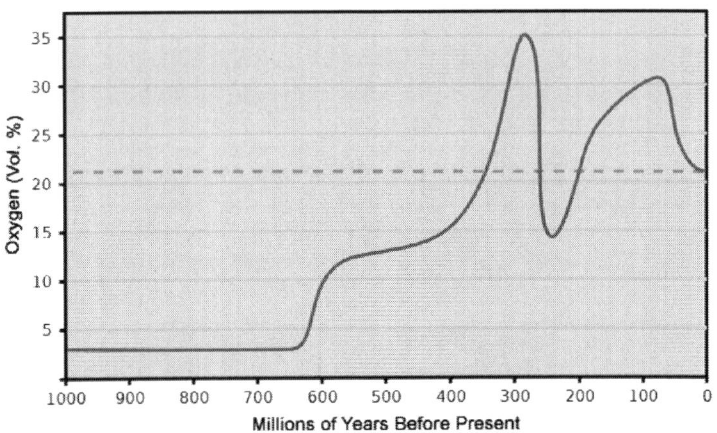

Figure: *Oxygen content of the atmosphere over the last billion years*

Currently, anthropogenic greenhouse gases are increasing in the atmosphere. According to the Intergovernmental Panel on Climate Change, this increase is the main cause of global warming.

Air Pollution

Air pollution is the introduction of chemicals, particulate matter, or biological materials that cause harm or discomfort to organisms into the atmosphere. Stratospheric ozone depletion is believed to be caused by air pollution (chiefly from chlorofluorocarbons).

Chapter 7

Radar Signal Processing

Distance Measurement

Transit time : One way to measure the distance to an object is to transmit a short pulse of radio signal (electromagnetic radiation), and measure the time it takes for the reflection to return. The distance is one-half the product of the round trip time (because the signal has to travel to the target and then back to the receiver) and the speed of the signal. Since radio waves travel at the speed of light (186,000 miles per second or 300,000,000 meters per second), accurate distance measurement requires high-performance electronics.

In most cases, the receiver does not detect the return while the signal is being transmitted. Through the use of a device called a *duplexer*, the radar switches between transmitting and receiving at a predetermined rate. The minimum range is calculated by measuring the length of the pulse multiplied by the speed of light, divided by two. In order to detect closer targets one must use a shorter pulse length.

A similar effect imposes a maximum range as well. If the return from the target comes in when the next pulse is being sent out, once again the receiver cannot tell the difference. In order to maximize range, longer times between pulses should be used, referred to as a pulse repetition time (PRT), or its reciprocal, pulse repetition frequency (PRF).

These two effects tend to be at odds with each other, and it is not easy to combine both good short range and good long range in a single radar. This is because the short pulses needed for a good minimum range broadcast have less total energy, making the returns much smaller and the target harder to detect. This could be offset by using more pulses, but this would shorten the maximum range again. So each radar uses a particular type of signal. Long-range radars tend to use long pulses with long delays between them, and short range radars use smaller pulses with less time between them. This pattern of pulses and pauses is known as the pulse repetition frequency (or PRF), and is one of the main ways to characterize a radar. As electronics have improved many radars now can change their PRF thereby changing their range. The newest radars fire 2 pulses during one cell, one for short range 10 km/6 miles and a separate signal for longer ranges 100 km/60 miles.

The distance resolution and the characteristics of the received signal as compared to noise depends heavily on the shape of the pulse. The pulse is often modulated to achieve better performance using a technique known as pulse compression.

Distance may also be measured as a function of time. The radar mile is the amount of time it takes for a radar pulse to travel one nautical mile, reflect off a target, and return to the radar antenna. Since a nautical mile is defined as *exactly* 1,852 meters, then dividing this distance by the speed of light (*exactly* 299,792,458 meters per second), and then multiplying the result by 2 (round trip = twice the distance), yields a result of approximately 12.36 microseconds in duration.

Frequency Modulation

Another form of distance measuring radar is based on frequency modulation. Frequency comparison between two signals is considerably more accurate, even with older electronics, than timing the signal. By measuring the frequency of the returned signal and comparing that with the original, the difference can be easily measured.

This technique can be used in continuous wave radar, and is often found in aircraft radar altimeters. In these systems a "carrier" radar signal is frequency modulated in a predictable way, typically varying up and down with a sine wave or sawtooth pattern at audio frequencies. The signal is then sent out from one antenna and received on another, typically located on the bottom of the aircraft, and the signal can be continuously compared using a simple *beat frequency* modulator that produces an audio frequency tone from the returned signal and a portion of the transmitted signal.

Since the signal frequency is changing, by the time the signal returns to the aircraft the broadcast has shifted to some other frequency. The amount of that shift is greater over longer times, so greater frequency differences mean a longer distance, the exact amount being the "ramp speed" selected by the electronics. The amount of shift is therefore directly related to the distance travelled, and can be displayed on an instrument. This signal processing is similar to that used in speed detecting Doppler radar. Example systems using this approach are AZUSA, MISTRAM, and UDOP.

A further advantage is that the radar can operate effectively at relatively low frequencies, comparable to that used by UHF television. This was important in the early development of this type when high frequency signal generation was difficult or expensive.

A new terrestrial radar uses low-power FM signals that cover a larger frequency range. The multiple reflections are analysed mathematically for pattern changes with multiple passes creating a computerized synthetic image. Doppler effects are not used which allows slow moving objects to be detected as well as largely eliminating "noise" from the surfaces of bodies of water. Used primarily for detection of intruders approaching in small boats or intruders crawling on the ground toward an objective.

Speed Measurement

Speed is the change in distance to an object with respect to time. Thus the existing system for measuring distance, combined with a memory capacity to see where the target last was, is enough to measure speed. At one time the memory consisted of a user making grease-pencil marks on the radar screen, and then calculating the speed using a slide rule. Modern radar systems perform the equivalent operation faster and more accurately using computers.

However, if the transmitter's output is coherent (phase synchronized), there is another effect that can be used to make almost instant speed measurements (no memory is required), known as the Doppler effect. Most modern radar systems use this principle in the pulse-doppler radar system. Return signals from targets are shifted away from this base frequency via the Doppler effect enabling the calculation of the speed of the object relative to the radar.

The Doppler effect is only able to determine the relative speed of the target along the line of sight from the radar to the target. Any component of target velocity perpendicular to the line of sight cannot be determined by using the Doppler effect alone, but it can be determined by tracking the target's azimuth over time. Additional information of the nature of the Doppler returns may be found in the radar signal characteristics article.

It is also possible to make a radar without any pulsing, known as a continuous-wave radar (CW radar), by sending out a very pure signal of a known frequency. CW radar is ideal for determining the radial component of a target's velocity, but it cannot determine the target's range. CW radar is typically used by traffic enforcement to measure vehicle speed quickly and accurately where range is not important.

Other mathematical developments in radar signal processing include time-frequency analysis (Weyl Heisenberg or wavelet), as well as the chirplet transform which makes use of the fact that radar returns from moving targets typically "chirp" (change their frequency as a function of time, as does the sound of a bird or bat).

Reduction of Interference Effects

Signal processing is employed in radar systems to reduce the radar interference effects. Signal processing techniques include moving target indication (MTI), pulse doppler, moving target detection (MTD) processors, correlation with secondary surveillance radar (SSR) targets, space-time adaptive processing (STAP), and track-before-detect (TBD). Constant false alarm rate (CFAR) and digital terrain model (DTM) processing are also used in clutter environments.

Plot and Track Extraction

Radar video returns on aircraft can be subjected to a plot extraction process whereby spurious and interfering signals are discarded. A sequence of target returns can be monitored through a device known as a plot extractor. The non relevant real time returns can be removed from the displayed information and a single plot displayed. In some radar systems, or alternatively in the command and control system to which the radar is connected, a radar tracker is used to associate the sequence of plots belonging to individual targets and estimate the targets' headings and speeds.

Radar Engineering

A radars components are:

- A transmitter that generates the radio signal with an oscillator such as a klystron or a magnetron and controls its duration by a modulator.
- A waveguide that links the transmitter and the antenna.
- A duplexer that serves as a switch between the antenna and the transmitter or the receiver for the signal when the antenna is used in both situations.

- A receiver. Knowing the shape of the desired received signal (a pulse), an optimal receiver can be designed using a matched filter.
- An electronic section that controls all those devices and the antenna to perform the radar scan ordered by a software.
- A link to end users.

Antenna Design

Radio signals broadcast from a single antenna will spread out in all directions, and likewise a single antenna will receive signals equally from all directions. This leaves the radar with the problem of deciding where the target object is located.

Early systems tended to use omni-directional broadcast antennas, with directional receiver antennas which were pointed in various directions. For instance the first system to be deployed, Chain Home, used two straight antennas at right angles for reception, each on a different display. The maximum return would be detected with an antenna at right angles to the target, and a minimum with the antenna pointed directly at it (end on). The operator could determine the direction to a target by rotating the antenna so one display showed a maximum while the other shows a minimum. One serious limitation with this type of solution is that the broadcast is sent out in all directions, so the amount of energy in the direction being examined is a small part of that transmitted. To get a reasonable amount of power on the "target", the transmitting aerial should also be directional.

Parabolic Reflector

More modern systems use a steerable parabolic "dish" to create a tight broadcast beam, typically using the same dish as the receiver. Such systems often combine two radar frequencies in the same antenna in order to allow automatic steering, or radar lock.

Parabolic reflectors can be either symmetric parabolas or spoiled parabolas:

- Symmetric parabolic antennas produce a narrow "pencil" beam in both the X and Y dimensions and consequently have a higher gain. The NEXRAD Pulse-Doppler weather radar uses a symmetric antenna to perform detailed volumetric scans of the atmosphere.
- Spoiled parabolic antennas produce a narrow beam in one dimension and a relatively wide beam in the other. This feature is useful if target detection over a wide range of angles is more important than target location in three dimensions. Most 2D

surveillance radars use a spoiled parabolic antenna with a narrow azimuthal beamwidth and wide vertical beamwidth. This beam configuration allows the radar operator to detect an aircraft at a specific azimuth but at an indeterminate height. Conversely, so-called "nodder" height finding radars use a dish with a narrow vertical beamwidth and wide azimuthal beamwidth to detect an aircraft at a specific height but with low azimuthal precision.

Types of Scan

- Primary Scan: A scanning technique where the main antenna aerial is moved to produce a scanning beam, examples include circular scan, sector scan etc.
- Secondary Scan: A scanning technique where the antenna feed is moved to produce a scanning beam, examples include conical scan, unidirectional sector scan, lobe switching etc.
- Palmer Scan: A scanning technique that produces a scanning beam by moving the main antenna and its feed. A Palmer Scan is a combination of a Primary Scan and a Secondary Scan.

Slotted Waveguide

Applied similarly to the parabolic reflector, the slotted waveguide is moved mechanically to scan and is particularly suitable for non-tracking surface scan systems, where the vertical pattern may remain constant. Owing to its lower cost and less wind exposure, shipboard, airport surface, and harbour surveillance radars now use this in preference to the parabolic antenna.

Phased Array

Another method of steering is used in a phased array radar. This uses an array of similar aerials suitably spaced, the phase of the signal to each individual aerial being controlled so that the signal is reinforced in the desired direction and cancels in other directions. If the individual aerials are in one plane and the signal is fed to each aerial in phase with all others then the signal will reinforce in a direction perpendicular to that plane. By altering the relative phase of the signal fed to each aerial the direction of the beam can be moved because the direction of constructive interference will move. Because phased array radars require no physical movement the beam can scan at thousands of degrees per second, fast enough to irradiate and track many individual targets, and still run a wide-ranging search periodically. By simply turning some of the antennas on or off, the beam can be spread for searching, narrowed for tracking, or even split into two or more virtual radars. However, the beam cannot be effectively steered at small angles to the plane of the array, so for full coverage multiple arrays are required, typically disposed on the faces of a triangular pyramid.

Phased array radars have been in use since the earliest years of radar use in World War II, but limitations of the electronics led to fairly poor accuracy. Phased array radars were originally used for missile defence. They are the heart of the ship-borne Aegis combat system, and the Patriot Missile System, and are increasingly used in other areas because the lack of moving parts makes them more reliable, and sometimes permits a much larger effective antenna, useful in fighter aircraft applications that offer only confined space for mechanical scanning.

As the price of electronics has fallen, phased array radars have become more and more common. Almost all modern military radar systems are based on phased arrays, where the small additional cost is far offset by the improved reliability of a system with no moving parts. Traditional moving-antenna designs are still widely used in roles where cost is a significant factor such as air traffic surveillance, weather radars and similar systems.

Phased array radars are also valued for use in aircraft, since they can track multiple targets. The first aircraft to use a phased array radar is the B-1B Lancer. The first aircraft fighter to use phased array radar was the Mikoyan MiG-31. The MiG-31M's SBI-16 Zaslon phased array radar is considered to be the world's most powerful fighter radar.

Phased-array interferometry or, aperture synthesis techniques, using an array of separate dishes that are phased into a single effective aperture, are not typically used for radar applications, although they are widely used in radio astronomy. Because of the Thinned array curse, such arrays of multiple apertures, when used in transmitters, result in narrow beams at the expense of reducing the total power transmitted to the target. In principle, such techniques used could increase the spatial resolution, but the lower power means that this is generally not effective. Aperture synthesis by post-processing of motion data from a single moving source, on the other hand, is widely used in space and airborne radar systems.

Radar Modulators

Modulators act to provide the waveform of the RF-pulse. There are two different radar modulator designs:

- high voltage switch for non-coherent keyed power-oscillators. These modulators consist of a high voltage pulse generator formed from a high voltage supply, a pulse forming network, and a high voltage switch such as a thyratron. They generate short pulses of power to feed the e.g. magnetron, a special type of vacuum tube that converts DC (usually pulsed) into microwaves. This technology is known as Pulsed power. In this way, the transmitted pulse of RF radiation is kept to a defined, and usually, very short duration.

- hybrid mixers, fed by a waveform generator and an exciter for a complex but coherent waveform. This waveform can be generated by low power/low-voltage input signals. In this case the radar transmitter must be a power-amplifier, e.g. a klystron tube or a solid state transmitter. In this way, the transmitted pulse is intrapulse modulated and the radar receiver must use pulse compression technique mostly.

Radar Coolant

Coolanol and PAO (poly-alpha olefin) are the two main coolants used to cool airborne radar equipment today.

Coolanol (silicate ester) was used in several military radars in the 1970s, for example the AN/APG-63 in the F-15. However, it is hygroscopic, leading to formation of highly flammable alcohol. The loss of a U.S. Navy aircraft in 1978 was attributed to a silicate ester fire. Coolanol is also expensive and toxic. The U.S. Navy has instituted a program named Pollution Prevention (P2) to reduce or eliminate the volume and toxicity of waste, air emissions, and effluent discharges. Because of this Coolanol is used less often today.

PAO is a synthetic lubricant blend of a polyol ester admixed with effective amounts of an antioxidant, yellow metal pacifier and rust inhibitors. The polyol ester blend includes a major proportion of poly (neopentyl polyol) ester blend formed by reacting poly (pentaerythritol) partial esters with at least one C7 to C12 carboxylic acid mixed with an ester formed by reacting a polyol having at least two hydroxyl groups and at least one C8-C10 carboxylic acid.

Preferably, the acids are linear and avoid those which can cause odours during use. Effective additives include secondary arylamine antioxidants, triazole derivative yellow metal pacifier and an amino acid derivative and substituted primary and secondary amine and/or diamine rust inhibitor.

A synthetic coolant/lubricant composition, comprising an ester mixture of 50 to 80 weight percent of poly (neopentyl polyol) ester formed by reacting a poly (neopentyl polyol) partial ester and at least one linear monocarboxylic acid having from 6 to 12 carbon atoms, and 20 to 50 weight percent of a polyol ester formed by reacting a polyol having 5 to 8 carbon atoms and at least two hydroxyl groups with at least one linear monocarboxylic acid having from 7 to 12 carbon atoms, the weight percents based on the total weight of the composition.

Radar Configurations and Types

Radars configurations include Monopulse radar, Bistatic radar, Doppler radar, Continuous-wave radar, etc.. depending on the types of hardware and software used. It is used in aviation (Primary and secondary radar), sea vessels, law enforcement, weather surveillance, ground mapping, geophysical surveys, and biological research.

Interferometric Synthetic Aperture Radar

Interferometric synthetic aperture radar, also abbreviated InSAR or IfSAR, is a radar technique used in geodesy and remote sensing. This geodetic method uses two or more synthetic aperture radar (SAR) images to generate maps of surface deformation or digital elevation, using differences in the phase of the waves returning to the satellite, or aircraft. The technique can potentially measure centimetre-scale changes in deformation over timespans of days to years. It has applications for geophysical monitoring of natural hazards, for example earthquakes, volcanoes and landslides, and also in structural engineering, in particular monitoring of subsidence and structural stability.

Synthetic Aperture Radar

Synthetic-aperture radar (SAR) is a form of radar whose defining characteristic is its use of relative motion between an antenna and its target region to provide distinctive long-term coherent-signal variations that are exploited to obtain finer spatial resolution than is possible with conventional beam-scanning means. It originated as an advanced form of side-looking airborne radar (SLAR).

SAR is usually implemented by mounting, on a moving platform such as an aircraft or spacecraft, a single beam-forming antenna from which a target scene is repeatedly illuminated with pulses of radio waves at wavelengths anywhere from a meter down to millimetres. The many echo waveforms received successively at the different antenna positions are coherently detected and stored and then post-processed together to resolve elements in an image of the target region.

Current (2010) airborne systems provide resolutions to about 10 cm, ultra-wideband systems provide resolutions of a few millimetres, and experimental terahertz SAR has provided sub-millimeter resolution in the laboratory. SAR images have wide applications in remote sensing and mapping of the surfaces of both the Earth and other planets. SAR can also be implemented as "inverse SAR" by observing a moving target over a substantial time with a stationary antenna.

Relationship to Phased Arrays

The term "synthetic", as used here, needs some clarification. That word came into popular use after chemists began to synthesize (literally, to assemble from subsidiary elements) materials sometimes not found in nature. It then began to be used to denote things thought of as "imitations" or even "fake".

A technique closely related to SAR uses an array (referred to as a "phased array") of real antenna elements spatially distributed over either one or two dimensions across the range dimension. These physical arrays are truly synthetic ones, indeed being created by synthesis of a collection of subsidiary physical antennas. Their operation need not involve motion relative to targets. All elements of these arrays receive simultaneously in real time, and the signals passing through them can be individually subjected to controlled shifts of the phases of those signals. One result can be to respond most strongly to radiation received from a specific small scene area, focusing on that area to determine its contribution to the total signal received. The coherently detected set of signals received over the entire array aperture can be replicated in several data-processing channels and processed differently in each. The set of responses thus traced to different small scene areas can be displayed together as an image of the scene.

In comparison, a SAR's (commonly) single physical antenna element gathers signals at different positions at different times. When the radar is carried by an aircraft or an orbiting vehicle, those positions are functions of a single variable, distance along the vehicle's path, which is a single mathematical dimension (not necessarily the same as a linear geometric dimension). The signals are stored, thus becoming functions, no longer of time, but of recording locations along that dimension. When the stored signals are read out later and combined with specific phase shifts, the result is the same as if the recorded data had been gathered by an equally long and shaped phased array. What is thus synthesized is a set of signals equivalent to what could have been received simultaneously by such an actual large-aperture (in one dimension) phased array. The SAR simulates (rather than synthesizes) that long one-dimensional phased array. Although the term in the title of this article has thus been incorrectly derived, it is now firmly established by half a century of usage.

While operation of a phased array is readily understood as a completely geometric technique, the fact that a synthetic aperture system gathers its data as it (or its target) moves at some speed means that phases which varied with the distance travelled originally varied with time, hence constituted temporal frequencies. Temporal frequencies being the variables commonly used by radar engineers, their analyses of SAR systems are usually (and very productively) couched in such terms. In particular, the variation of phase during flight over the length of the synthetic aperture is seen as a sequence of Doppler shifts of the received frequency from that of the transmitted frequency. It is significant, though, to realize that, once the received data have been recorded and thus have become timeless, the SAR data-processing situation is also understandable as a special type of phased array, treatable as a completely geometric process.

The core of both the SAR and the phased array techniques is that the distances that radar waves travel to and back from each scene element consist of some integer number of wavelengths plus some fraction of a "final" wavelength. Those fractions cause differences between the phases of the re-radiation received at various SAR or array positions. Coherent detection is needed to capture the signal phase information in addition to the signal amplitude information. That type of detection requires finding the differences between the phases of the received signals and the simultaneous phase of a well-preserved sample of the transmitted illumination.

Every wave scattered from any point in the scene has a circular curvature about that point as a centre. Signals from scene points at different ranges therefore arrive at a planar array with different curvatures, resulting in signal phase changes which follow different quadratic variations across a planar phased array. Additional linear variations result from points located in different directions from the centre of the array. Fortunately, any one combination of these variations is unique to one scene point, and is calculable. For a SAR, the two-way travel doubles that phase change.

In reading the following two paragraphs, be particularly careful to distinguish between array elements and scene elements. Also remember that each of the latter has, of course, a matching image element.

Comparison of the array-signal phase variation across the array with the total calculated phase variation pattern can reveal the relative portion of the total received signal that came from the only scene point that could be responsible for that pattern. One way to do the comparison is by a correlation computation, multiplying, for each scene element, the received and the calculated field-intensity values array element by array element and then summing the products for each scene element. Alternatively, one could, for each scene element, subtract each array element's calculated phase shift from the actual received phase and then vectorially sum the resulting field-intensity differences over the array. Wherever in the scene the two phases substantially cancel everywhere in the array, the difference vectors being added are in phase, yielding, for that scene point, a maximum value for the sum.

The equivalence of these two methods can be seen by recognizing that multiplication of sinusiods can be done by summing phases which are complex-number exponents of e, the base of natural logarithms.

However it is done, the image-deriving process amounts to "backtracking" the process by which nature previously spread the scene information over the array. In each direction, the process may be viewed as a Fourier transform, which is a type of correlation process. The image-extraction process we use can then be seen as another Fourier transform which is a reversal of the original natural one.

It is important to realize that only those sub-wavelength differences of successive ranges from the transmitting antenna to each target point and back, which govern signal phase, are used to refine the resolution in any geometric dimension. The central direction and the angular width of the illuminating beam do not contribute directly to creating that fine resolution. Instead, they serve only to select the solid-angle region from which usable range data are received. While some distinguishing of the ranges of different scene items can be made from the forms of their sub-wavelength range variations at short ranges, the very large depth of focus that occurs at long ranges usually requires that over-all range differences (larger than a wavelength) be used to define range resolutions comparable to the achievable cross-range resolution.

Typical Operation

In a typical SAR application, a single radar antenna is attached to an aircraft or spacecraft so as to radiate a beam whose wave-propagation direction has a substantial component perpendicular to the flight-path direction. The beam is allowed to be broad in the vertical direction so it will illuminate the terrain from nearly beneath the aircraft out toward the horizon.

Resolution in the range dimension of the image is accomplished by creating pulses which define very short time intervals, either by emitting short pulses consisting of a "carrier" frequency and the necessary "sidebands", all within a certain bandwidth, or by using longer "chirp pulses" in which frequency varies (often linearly) with time within that bandwidth. The differing times at which echoes return allow points at different distances to be distinguished.

The total signal is that from a beamwidth-sized patch of ground. To produce a beam that is narrow in the cross-range direction, diffraction effects require that the antenna be wide in that dimension. Therefore the distinguishing, from each other, of co-range points simply by strengths of returns that persist for as long as they are within the beam width is difficult with aircraft-carryable antennas, because their beams can have linear widths only about two orders of magnitude (hundreds of times) smaller than the range. (Spacecraft-carryable ones can do 10 or more times better.) However, if both the amplitude and the phase of returns are recorded, then the portion of that multi-target return that was scattered radially from any smaller scene element can be extracted by phase-vector correlation of the total return with the form of the return expected from each such element. Careful design and operation can accomplish resolution of items smaller than a millionth of the range, for example, 30 cm at 300 km, or about one foot at nearly 200 miles.

The process can be thought of as combining the series of spatially distributed observations as if all had been made simultaneously with an antenna as long as the beamwidth and focused on that particular point. The "synthetic aperture" simulated at maximum system range by this process not only is longer than the real antenna, but, in practical applications, it is much longer than the radar aircraft, and tremendously longer than the radar spacecraft.

Image resolution of SAR in its range coordinate (expressed in image pixels per distance unit) is mainly proportional to the radio bandwidth of whatever type of pulse is used. In the cross-range coordinate, the similar resolution is mainly proportional to the bandwidth of the Doppler shift of the signal returns within the beamwidth. Since Doppler frequency depends on the angle of the scattering point's direction from the broadside direction, the Doppler bandwidth available within the beamwidth is the same at all ranges. Hence the theoretical spatial resolution limits in both image dimensions remain constant with variation of range. However, in practice, both the errors that accumulate with data-collection time and the particular techniques used in post-processing further limit cross-range resolution at long ranges.

The conversion of return delay time to geometric range can be very accurate because of the natural constancy of the speed and direction of propagation of electromagnetic waves. However, for an aircraft flying through the never-uniform and never-quiescent atmosphere, the relating of pulse transmission and reception times to successive geometric positions of the antenna must be accompanied by constant adjusting of the return phases to account for sensed irregularities in the flight path. SAR's in spacecraft avoid that atmosphere problem, but still must make corrections for known antenna movements due to rotations of the spacecraft, even those that are reactions to movements of onboard machinery. Locating a SAR in a manned space vehicle may require that the humans carefully remain motionless relative to the vehicle during data collection periods.

Although some references to SARs have characterized them as "radar telescopes", their actual optical analogy is the microscope, the detail in their images being smaller than the length of the synthetic aperture. In radar-engineering terms, while the target area is in the "far field" of the illuminating antenna, it is in the "near field" of the simulated one.

Returns from scatterers within the range extent of any image are spread over a matching time interval. The inter-pulse period must be long enough to allow farthest-range returns from any pulse to finish arriving before the nearest-range ones from the next pulse begin to appear, so that those do not overlap each other in time. On the other hand, the interpulse rate must be fast enough to provide sufficient samples for the desired across-range (or across-beam) resolution. When the radar is to be carried by a high-speed vehicle and is to image a large area at fine resolution, those conditions may clash, leading to what has been called SAR's ambiguity problem. The same considerations apply to "conventional" radars also, but this problem occurs significantly only when resolution is so fine as to be available only through SAR processes. Since the basis of the problem is the information-carrying capacity of the single signal-input channel provided by one antenna, the only solution is to use additional channels fed by additional antennas. The system then becomes a hybrid of a SAR and a phased array, sometimes being called a Vernier Array.

Combining the series of observations requires significant computational resources, usually using Fourier transform techniques. The high digital computing speed now available allows such processing to be done in near-real time on board a SAR aircraft. (There is necessarily a minimum time delay until all parts of the signal have been received.) The result is a map of radar reflectivity, including both amplitude and phase. The amplitude information, when shown in a map-like display, gives information about ground cover in much the same way that a black-and-white photo does. Variations in processing may also be done in either vehicle-borne stations or ground stations for various purposes, so as to accentuate certain image features for detailed target-area analysis.

Although the phase information in an image is generally not made available to a human observer of an image display device, it can be preserved numerically, and sometimes allows certain additional features of targets to be recognized. Unfortunately, the phase differences between adjacent image picture elements ("pixels") also produce random interference effects called "coherence speckle", which is a sort of graininess with dimensions on the order of the resolution, causing the concept of resolution to take on a subtly different meaning. This effect is the same as is apparent both visually and photographically in laser-illuminated optical scenes. The scale of that random speckle structure is governed by the size of the synthetic aperture in wavelengths, and cannot be finer than the system's resolution. Speckle structure can be subdued at the expense of resolution.

Before rapid digital computers were available, the data processing was done using an optical holography technique. The analog radar data were recorded as a holographic interference pattern on photographic film at a scale permitting the film to preserve the signal bandwidths (for example, 1:1,000,000 for a radar using a 0.6-meter wavelength). Then light using, for example, 0.6-micrometer waves (as from a helium-neon laser) passing through the hologram could project a terrain image at a scale recordable on another film at reasonable processor focal distances of around a meter. This worked because both SAR and phased arrays are fundamentally similar to optical holography, but using microwaves instead of light waves. The "optical data-processors" developed for this radar purpose were the first effective analog optical computer systems, and were, in fact, devised before the holographic technique was fully adapted to optical imaging. Because of the different sources of range and across-range signal structures in the radar signals, optical data-processors for SAR included not only both spherical and cylindrical lenses, but sometimes conical ones.

Image Appearance

The following considerations apply also to real-aperture terrain-imaging radars, but are more consequential when resolution in range is matched to a cross-beam resolution that is available only from a SAR.

The two dimensions of a radar image are range and cross-range. Radar images of limited patches of terrain can resemble oblique photographs, but not ones taken from the location of the radar. This is because the range coordinate in a radar image is perpendicular to the vertical-angle coordinate of an oblique photo. The apparent entrance-pupil position (or camera centre) for viewing such an image is therefore not as if at the radar, but as if at a point from which the viewer's line of sight is perpendicular to the slant-range direction connecting radar and target, with slant-range increasing from top to bottom of the image.

Because slant ranges to level terrain vary in vertical angle, each elevation of such terrain appears as a curved surface, specifically a hyperbolic cosine one. Verticals at various ranges are perpendiculars to those curves. The viewer's apparent looking directions are parallel to the curve's "hypcos" axis. Items directly beneath the radar appear as if optically viewed horizontally (i.e., from the side) and those at far ranges as if optically viewed from directly above. These curvatures are not evident unless large extents of near-range terrain, including steep slant ranges, are being viewed.

When viewed as specified above, fine-resolution radar images of small areas can appear most nearly like familiar optical ones, for two reasons. The first reason is easily understood by imagining a flagpole in the scene. The slant-range to its upper end is less than that to its base. Therefore the pole can appear correctly top-end up only when viewed in the above orientation. Secondly, the radar illumination then being downward, shadows are seen in their most-familiar "overhead-lighting" direction.

Note that the image of the pole's top will overlay that of some terrain point which is on the same slant range arc but at a shorter horizontal range ("ground-range"). Images of scene surfaces which faced both the illumination and the apparent eyepoint will have geometries that resemble those of an optical scene viewed from that eyepoint. However, slopes facing the radar will be foreshortened and ones facing away from it will be lengthened from their horizontal (map) dimensions. The former will therefore be brightened and the latter dimmed.

Returns from slopes steeper than perpendicular to slant range will be overlaid on those of lower-elevation terrain at a nearer ground-range, both being visible but intermingled. This is especially the case for vertical surfaces like the walls of buildings. Another viewing inconvenience that arises when a surface is steeper than perpendicular to the slant range is that it is then illuminated on one face but "viewed" from the reverse face. Then one "sees", for example, the radar-facing wall of a building as if from the inside, while the building's interior and the rear wall (that nearest to, hence expected to be optically visible to, the viewer) have vanished, since they lack illumination, being in the shadow of the front wall and the roof. Some return from the roof may overlay that from the front wall, and both of those may overlay return from terrain in front of the building. The visible building shadow will include those of all illuminated items. Long shadows may exhibit blurred edges due to the illuminating antenna's movement during the "time exposure" needed to create the image.

Surfaces that we usually consider rough will, if that roughness consists of relief less than the radar wavelength, behave as smooth mirrors, showing, beyond such a surface, additional images of items in front of it. Those mirror images will appear within the shadow of the mirroring surface, sometimes filling the entire shadow, thus preventing recognition of the shadow.

An important fact that applies to SARs but not to real-aperture radars is that the direction of overlay of any scene point is not directly toward the radar, but toward that point of the SAR's current path direction that is nearest to the target point. If the SAR is "squinting" forward or aft away from the exactly broadside direction, then the illumination direction, and hence the shadow direction, will not be opposite to the overlay direction, but slanted to right or left from it. An image will appear with the correct projection geometry when viewed so that the overlay direction is vertical, the SAR's flight-path is above the image, and range increases somewhat downward.

Objects in motion within a SAR scene alter the Doppler frequencies of the returns. Such objects therefore appear in the image at locations offset in the across-range direction by amounts proportional to the range-direction component of their velocity. Road vehicles may be depicted off the roadway and therefore not recognized as road traffic items.

Trains appearing away from their tracks are more easily properly recognized by their length parallel to known trackage as well as by the absence of an equal length of railbed signature and of some adjacent terrain, both having been shadowed by the train. While images of moving vessels can be offset from the line of the earlier parts of their wakes, the more recent parts of the wake, which still partake of some of the vessel's motion, appear as curves connecting the vessel image to the relatively quiescent far-aft wake. In such identifiable cases, speed and direction of the moving items can be determined from the amounts of their offsets. The along-track component of a target's motion causes some defocus. Random motions such as that of wind-driven tree foliage, vehicles driven over rough terrain, or humans or other animals walking or running generally render those items not focusable, resulting in blurring or even effective invisibility.

These considerations, along with the speckle structure due to coherence, take some getting used to in order to correctly interpret SAR images. To assist in that, large collections of significant target signatures have been accumulated by performing many test flights over known terrains and cultural objects.

Origin and Early Development (ca. 1950-1975)

Carl A. Wiley, a mathematician at Goodyear Aircraft Corporation in Litchfield Park, Arizona, invented synthetic-aperture radar in June 1951 while working on a correlation guidance system for the *Atlas* ICBM program. In early 1952, Wiley, together with Fred Heisley and Bill Welty, constructed a concept validation system known as DOUSER ("Doppler Unbeamed Search Radar"). During the 1950s and 1960s Goodyear Aircraft (later Goodyear Aerospace) introduced numerous advancements in SAR technology.

Independently of Wiley's work, experimental trials in early 1952 by Sherwin and others at the University of Illinois' Control Systems Laboratory showed results that they pointed out "could provide the basis for radar systems with greatly improved angular resolution" and might even lead to systems capable of focusing at all ranges simultaneously.

In both of those programs, processing of the radar returns was done by electrical-circuit filtering methods. In essence, signal strength in isolated discrete bands of Doppler frequency defined image intensities that were displayed at matching angular positions within proper range locations. When only the central (zero-Doppler band) portion of the return signals was used, the effect was as if only that central part of the beam existed.

That led to the term Doppler Beam Sharpening. Displaying returns from several adjacent non-zero Doppler frequency bands accomplished further "beam-subdividing" (sometimes called "unfocused radar", though it could have been considered "semi-focused"). Wiley's patent, applied for in 1954, still proposed similar processing. The bulkiness of the circuitry then available limited the extent to which those schemes might further improve resolution.

The principle was included in a memorandum authored by Walter Hausz of General Electric that was part of the then-secret report of a

1952 Dept. of Defence summer study conference called TEOTA ("The Eyes of the Army"), which sought to identify new techniques useful for military reconnaissance and technical gathering of intelligence. A follow-on summer program in 1953 at The University of Michigan, called Project Wolverine, identified several of the TEOTA subjects, including Doppler-assisted sub-beamwidth resolution, as research efforts to be sponsored by the Dept. of Defence at various academic and industrial research laboratories. In that same year, the Illinois group produced a "strip-map" image exhibiting a considerable amount of sub-beamwidth resolution.

A more advanced focused-radar project was among several remote sensing schemes assigned in 1953 to Project Michigan, a tri-service-sponsored (Army, Navy, Air Force) program at the University of Michigan's Willow Run Research Centre (WRRC), that program being administered by the Army Signal Corps.

Initially called the side-looking radar project, it was carried out by a group first known as the Radar Laboratory and later as the Radar and Optics Laboratory. It proposed to take into account, not just the short-term existence of several particular Doppler shifts, but the entire history of the steadily varying shifts from each target as the latter crossed the beam. An early analysis by Dr. Louis J. Cutrona, Weston E. Vivian, and Emmett N. Leith of that group showed that such a fully focused system should yield, at all ranges, a resolution equal to the width (or, by some criteria, the half-width) of the real antenna carried on the radar aircraft and continually pointed broadside to the aircraft's path.

The required data processing amounted to calculating cross-correlations of the received signals with samples of the forms of signals to be expected from unit-amplitude sources at the various ranges. At that time, even "large" digital computers had capabilities somewhat near the levels of today's four-function hand calculators, hence were nowhere near able to do such a huge amount of computation. Instead, the device for doing the correlation computations was to be an optical correlator.

It was proposed that signals received by the travelling antenna and coherently detected be displayed as a single range-trace line across the diameter of the face of a cathode-ray tube, the line's successive forms being recorded as images projected onto a film travelling perpendicular to the length of that line. The information on the developed film was to be subsequently processed in the laboratory on equipment still to be devised as a principal task of the project.

In the initial processor proposal, an arrangement of lenses was expected to multiply the recorded signals point-by-point with the known signal forms by passing light successively through both the signal film and another film containing the known signal pattern. The subsequent summation, or integration, step of the correlation was to be done by converging appropriate sets of multiplication products by the focusing action of one or more spherical and cylindrical lenses. The processor was to be, in effect, an optical analog computer performing large-scale scalar-arithmetic calculations in many channels (with many light "rays") at once. Ultimately, two such devices would be needed, their outputs to be combined as quadrature components of the complete solution.

Fortunately (as it turned out), a desire to keep the equipment small had led to recording the reference pattern on 35-mm film. Trials promptly showed that the patterns on the film were so fine as to show pronounced diffraction effects that prevented sharp final focusing.

That led Leith, a physicist who was devising the correlator, to recognize that those effects in themselves could, by natural processes, perform a significant part of the needed processing, since along-track strips of the recording operated like diametrical slices of a series of circular optical zone plates. Any such plate performs somewhat like a lens, each plate having a specific focal length for any given wavelength. The recording that had been considered as scalar became recognized as pairs of opposite-sign vector ones of many spatial frequencies plus a zero-frequency "bias" quantity. The needed correlation summation changed from a pair of scalar ones to a single vector one.

Each zone plate strip has two equal but oppositely-signed focal lengths, one real, where a beam through it converges to a focus, and one virtual, where another beam appears to have diverged from, beyond the other face of the zone plate. The zero-frequency (d-c bias) component has no focal point, but overlays both the converging and diverging beams. The key to obtaining, from the converging wave component, focused images that are not overlaid with unwanted haze from the other two is to block the latter, allowing only the wanted beam to pass through a properly positioned frequency-band selecting aperture.

Each radar range yields a zone plate strip with a focal length proportional to that range. This fact became a principal complication in the design of optical processors. Consequently, technical journals of the time contain a large volume of material devoted to ways for coping with the variation of focus with range.

For that major change in approach, the light used had to be both monochromatic and coherent, properties that were already a requirement on the radar radiation. Lasers also then being in the future, the best then-available approximation to a coherent light source was the output of a mercury-vapour lamp, passed through a colour filter that was matched to the lamp spectrum's green band, and then concentrated as well as possible onto a very small beam-limiting aperture. While the resulting amount of light was so weak that very long exposure times had to be used, a workable optical correlator was assembled in time to be used when appropriate data became available.

Although creating that radar was a more straight-forward task based on already-known techniques, that work did demand the achievement of signal linearity and frequency stability that were at the extreme state of the art. An adequate instrument was designed and built by the Radar Laboratory and was installed in a C-46 (Curtiss "Commando") aircraft.

Because the aircraft was bailed to WRRC by the U. S. Army and was flown and maintained by WRRC's own pilots and ground personnel, it was available for many flights at times matching the Radar Laboratory's needs, a feature important for allowing frequent re-testing and "debugging" of the continually developing complex equipment. By contrast, the Illinois group had used a C-46 belonging to the Air Force and flown by AF pilots only by pre-arrangement, resulting, in the eyes of those researchers, in limitation to a less-than-desirable frequency of flight tests of their equipment, hence a low bandwidth of feedback from tests. (Later work with newer Convair aircraft continued the Michigan group's local control of flight schedules.)

Michigan's chosen 5-foot-wide WWII-surplus antenna was theoretically capable of 5-foot resolution, but data from only 10% of the beamwidth was used at first, the goal at that time being to demonstrate 50-foot resolution. It was understood that finer resolution would require the added development of means for sensing departures of the aircraft from an ideal heading and flight path, and for using that information for making needed corrections to the antenna pointing and to the received signals before processing.

After numerous trials in which even small atmospheric turbulence kept the aircraft from flying straight and level enough for good 50-foot data, one pre-dawn flight in August 1957. Yielded a map-like image of the Willow Run Airport area which did demonstrate 50-foot resolution in some parts of the image, whereas the illuminated beam width there was 900 feet. Although the program had been considered for termination by DoD due to what had seemed to be a lack of results, that first success ensured further funding to continue development leading to solutions to those recognized needs. The SAR principle was first acknowledged publicly via an April, 1960 press release about the U. S. Army experimental AN/UPD-1 system, which consisted of an airborne element made by Texas Instruments and installed in a Beech L-23D aircraft and a mobile ground data-processing station made by WRRC and installed in a military van. At the time, the nature of the data-processor was not revealed.

A technical article in the journal of the IRE (Institute of Radio Engineers) Professional Group on Military Electronics in February 1961 described the SAR principle and both the C-46 and AN/UPD-1 versions, but did not tell how the data were processed, nor that the UPD-1's maximum resolution capability was about 50 feet. However, the June, 1960 issue of the IRE Professional Group on Information Theory had contained a long article on "Optical Data Processing and Filtering Systems" by members of the Michigan group. Although it did not refer to the use of those techniques for radar, readers of both journals could quite easily understand the existence of a connection between articles sharing some authors.

An operational system to be carried in a reconnaissance version of the F-4 "Phantom" aircraft was quickly devised and was used briefly in Vietnam, where it failed to favourably impress its users, due to the combination of its low resolution (similar to the UPD-1's), the speckly nature of its coherent-wave images (similar to the speckliness of laser images), and the poorly understood dissimilarity of its range/cross-range images from the angle/angle optical ones familiar to military photo interpreters. The lessons it provided were well learned by subsequent researchers, operational system designers, image-interpreter trainers, and the DoD sponsors of further development and acquisition.

In subsequent work, the technique's latent capability was eventually achieved. That work, depending on advanced radar circuit designs and precision sensing of departures from ideal straight flight, along with more sophisticated optical processors using laser light sources and specially-designed very large lenses made from remarkably clear glass, allowed the Michigan group to advance system resolution, at about 5-year intervals, first to 15 feet, then 5 feet, and, by the mid-1970s, to 1 foot (the latter only over very short range intervals while processing was still being done optically).

The latter levels and the associated very wide dynamic range proved suitable for identifying many objects of military concern as well as soil, water, vegetation, and ice features being studied by a variety of environmental researchers having security clearances allowing them access to what was then classified imagery. Similarly improved operational systems soon followed each of those finer-resolution steps.

Even the 5-foot resolution stage had over-taxed the ability of cathode-ray tubes (limited to about 2000 distinguishable items across the screen diameter) to deliver fine enough details to signal films while still covering wide range swaths, and taxed the optical processing systems in similar ways. However, at about the same time, digital computers finally became capable of doing the processing without similar limitation, and the consequent presentation of the images on cathode ray tube monitors instead of film allowed for better control over tonal reproduction and for more convenient image mensuration.

Achievement of the finest resolutions at long ranges was aided by adding the capability to swing a larger airborne antenna so as to more strongly illuminate a limited target area continually while collecting data over several degrees of aspect, removing the previous limitation of resolution to the antenna width. This was referred to as the spotlight mode, which no longer produced continuous-swath images but, instead, images of isolated patches of terrain. It was understood very early in SAR development that the extremely smooth orbital path of an out-of-the-atmosphere platform made it ideally suited to SAR operation.

Early experience with artificial earth satellites had also demonstrated that the Doppler frequency shifts of signals travelling through the ionosphere and atmosphere were stable enough to permit very fine resolution to be achievable even at ranges of hundreds of kilometers. While experimental verification of those facts (still classified after nearly half a century) occurred within the second decade after the initial work began, several of the capabilities for creating useful classified systems did not exist for another two decades.

That seemingly slow rate of advances was often paced by the progress of other inventions, such as the laser, the digital computer, circuit miniaturization, and compact data storage. Once the laser appeared, optical data-processing became a fast process because it provided many parallel analog channels, but devising optical chains suited to matching signal focal lengths to ranges proceeded by many stages and turned out to call for some novel optical components. Since the process depended on diffraction of light waves, it required anti-vibration mountings, clean rooms, and highly trained operators. Even at its best, its use of CRT's and film for data storage placed limits on the range depth of images.

At several stages, attaining the frequently over-optimistic expectations for digital computation equipment proved to take far longer than anticipated. For example, the SEASAT system was ready to orbit before its digital processor became available, so a quickly assembled optical recording and processing scheme had to be used to obtain timely confirmation of system operation. When its digital processor was finally completed and used, the digital equipment of that time took many hours to create one swath of image from each run of a few seconds of data. Still, while that was a step down in speed, it was a step up in image quality. Modern methods now provide both high speed and high quality.

Although the above specifies the system development contributions of only a few organizations, many other groups had also become players as the value of SAR became more and more apparent. Especially crucial to the organization and funding of the initial long development process was the technical expertise and foresight of a number of both civilian and uniformed project managers in equipment procurement agencies in the federal government, particularly, of course, ones in the armed forces and in the intelligence agencies, and also in some civilian space agencies.

Algorithm

The SAR algorithm, as given here, applies to phased arrays generally.

A three-dimensional array (a volume) of scene elements is defined which will represent the volume of space within which targets exist. Each element of the array is a cubical voxel representing the probability (a "density") of a reflective surface being at that location in space. (Note that two-dimensional SARs are also possible—showing only a top-down view of the target area). Initially, the SAR algorithm gives each voxel a density of zero.

Then, for each captured waveform, the entire volume is iterated. For a given waveform and voxel, the distance from the position represented by that voxel to the antenna(e) used to capture that waveform is calculated. That distance represents a time delay into the waveform. The sample value at that position in the waveform is then added to the voxel's density value.

This represents a possible echo from a target at that position. Note that there are several optional approaches here, depending on the precision of the waveform timing, among other things. For example, if phase cannot be accurately known, then only the envelope magnitude (with the help of a Hilbert transform) of the waveform sample might be added to the voxel. If polarization and phase are known in the waveform, and are accurate enough, then these values might be added to a more complex voxel that holds such measurements separately.

After all waveforms have been iterated over all voxels, the basic SAR processing is complete.

What remains, in the simplest approach, is to decide what voxel density value represents a solid object. Voxels whose density is below that threshold are ignored. Note that the threshold level chosen must at least be higher than the peak energy of any single wave—otherwise that wave peak would appear as a sphere (or ellipse, in the case of multistatic operation) of false "density" across the entire volume. Thus to detect a point on a target, there must be at least two different antenna echoes from that point. Consequently, there is a need for large numbers of antenna positions to properly characterize a target.

The voxels that passed the threshold criteria are visualized in 2D or 3D. Optionally, added visual quality can sometimes be had by use of a surface detection algorithm like marching cubes.

More Complex Operation

The basic design of a synthetic-aperture radar system can be enhanced to collect more information. Most of these methods use the same basic principle of combining many pulses to form a synthetic aperture, but may involve additional antennae or significant additional processing.

Multistatic Operation

SAR requires that echo captures be taken at multiple antenna positions. The more captures taken (at different antenna locations) the more reliable the target characterization. Multiple captures can be obtained by moving a single antenna to different locations, by placing multiple stationary antennae at different locations, or combinations thereof. The advantage of a single moving antenna is that it can be easily placed in any number of positions to provide any number of monostatic waveforms. For example, an antenna mounted on an airplane takes many captures per second as the plane travels.

The principal advantages of multiple static antennae are that a moving target can be characterized (assuming the capture electronics are fast enough), that no vehicle or motion machinery is necessary, and that antenna positions need not be derived from other, sometimes unreliable, information. (One problem with SAR aboard an airplane is knowing precise antenna positions as the plane travels).

For multiple static antennae, all combinations of monostatic and multistatic radar waveform captures are possible. Note, however, that it is not advantageous to capture a waveform for each of both transmission directions for a given pair of antennae, because those waveforms will be identical. When multiple static antennae are used, the total number of unique echo waveforms that can be captured is

$$\frac{N^2 + N}{2}$$

Where N is the number of unique antenna positions.

Polarimetry

Radar waves have a polarization. Different materials reflect radar waves with different intensities, but anisotropic materials such as grass often reflect different polarizations with different intensities. Some materials will also convert one polarization into another. By emitting a mixture of polarizations and using receiving antennae with a specific polarization, several different images can be collected from the same series of pulses. Frequently three such RX-TX polarizations (HH-pol, VV-pol, VH-pol) are used as the three colour channels in a synthesized image. This is what has been done in the picture at left. Interpretation of the resulting colours requires significant testing of known materials.

New developments in polarimetry also include utilizing the changes in the random polarization returns of some surfaces (such as grass or sand), between two images of the same location at different points in time to determine where changes not visible to optical systems occurred. Examples include subterranean tunneling, or paths of vehicles driving through the area being imaged. Enhanced SAR sea oil slick observation it has been developed by appropriate physical modelling and use of fully-polarimetric and dual-polarimetric measurements.

Interferometry

Rather than discarding the phase data, information can be extracted from it. If two observations of the same terrain from very similar positions are available, aperture synthesis can be performed to provide the resolution performance which would be given by a RADAR system with dimensions equal to the separation of the two measurements. This technique is called Interferometric SAR or InSAR.

If the two samples are obtained simultaneously (perhaps by placing two antennas on the same aircraft, some distance apart), then any phase difference will contain information about the angle from which the radar echo returned. Combining this with the distance information, one can determine the position in three dimensions of the image pixel.

In other words, one can extract terrain altitude as well as radar reflectivity, producing a digital elevation model (DEM) with a single airplane pass. One aircraft application at the Canada Centre for Remote Sensing produced digital elevation maps with a resolution of 5 m and altitude errors also on the order of 5 m. Interferometry was used to map many regions of the Earth's surface with unprecedented accuracy using data from the Shuttle Radar Topography Mission. If the two samples are separated in time, perhaps from two different flights over the same terrain, then there are two possible sources of phase shift.

The first is terrain altitude, as discussed above. The second is terrain motion: if the terrain has shifted between observations, it will return a different phase. The amount of shift required to cause a significant phase difference is on the order of the wavelength used. This means that if the terrain shifts by *centimetres*, it can be seen in the resulting image (A digital elevation map must be available in order to separate the two kinds of phase difference; a third pass may be necessary in order to produce one).

This second method offers a powerful tool in geology and geography. Glacier flow can be mapped with two passes. Maps showing the land deformation after a minor earthquake or after a volcanic eruption (showing the shrinkage of the whole volcano by several centimetres) have been published.

Differential Interferometry

Differential interferometry (D-InSAR) requires taking at least two images with addition of a DEM. The DEM can be either produced by GPS measurements or could be generated by interferometry as long as the time between acquisition of the image pairs is short, which guarantees minimal distortion of the image of the target surface. In principle, 3 images of the ground area with similar image acquisition geometry is often adequate for D-InSar.

The principle for detecting ground movement is quite simple. One interferogram is created from the first two images; this is also called the reference interferogram or topographical interferogram. A second interferogram is created that captures topography + distortion. Subtracting the latter from the reference interferogram can reveal differential fringes, indicating movement. The described 3 image D-InSAR generation technique is called 3-pass or double-difference method.

Differential fringes which remain as fringes in the differential interferogram are a result of SAR range changes of any displaced point on the ground from one interferogram to the next. In the differential interferogram, each fringe is directly proportional to the SAR wavelength, which is about 5.6 cm for ERS and RADARSAT single phase cycle. Surface displacement away from the satellite look direction causes an increase in path (translating to phase) difference. Since the signal travels from the SAR antenna to target and back again, the measured displacement is twice the unit of wavelength. This means in differential interferometry one fringe cycle -pi to +pi or one wavelength corresponds to a displacement relative to SAR antenna of only half wavelength (2.8 cm). There are various publications on measuring subsidence movement, slope stability analysis, landslide, glacier movement, etc. tooling D-InSAR. Further advancement to this technique whereby differential interferometry from satellite SAR ascending pass and descending pass can be used to estimate 3-D ground movement. Research in this area has shown accurate measurements of 3-D ground movement with accuracies comparable to GPS based measurements can be achieved.

Ultra-wideband SAR

Conventional radar systems emit bursts of radio energy with a fairly narrow range of frequencies. A narrow-band channel, by definition, does not allow rapid changes in modulation. Since it is the change in a received signal that reveals the time of arrival of the signal (obviously an unchanging signal would reveal nothing about "when" it reflected from the target), a signal with only a slow change in modulation cannot reveal the distance to the target as well as can a signal with a quick change in modulation.

Ultra-wideband (UWB) refers to any radio transmission that uses a very large bandwidth – which is the same as saying it uses very rapid changes in modulation. Although there is no set bandwidth value that qualifies a signal as "UWB", systems using bandwidths greater than a sizable portion of the centre frequency (typically about ten percent, or so) are most often called "UWB" systems. A typical UWB system might use a bandwidth of one-third to one-half of its centre frequency. For example, some systems use a bandwidth of about 1 GHz centred around 3 GHz.

There are as many ways to increase the bandwidth of a signal as there are forms of modulation – it is simply a matter of increasing the rate of that modulation. However, the two most common methods used in UWB radar, including SAR, are very short pulses and high-bandwidth chirping. A general description of chirping appears elsewhere in this article. The bandwidth of a chirped system can be as narrow or as wide as the designers desire. Pulse-based UWB systems, being the more common method associated with the term "UWB radar", are described here.

A pulse-based radar system transmits very short pulses of electromagnetic energy, typically only a few waves or less. A very short pulse is, of course, a very rapidly changing signal, and thus occupies a very wide bandwidth. This allows far more accurate measurement of distance, and thus resolution.

The main disadvantage of pulse-based UWB SAR is that the transmitting and receiving front-end electronics are difficult to design for high-power applications. Specifically, the transmit duty cycle is so exceptionally low and pulse time so exceptionally short, that the electronics must be capable of extremely high instantaneous power to rival the average power of conventional radars. (Although it is true that UWB provides a notable gain in channel capacity over a narrow band signal because of the relationship of bandwidth in the Shannon–Hartley theorem and because the low receive duty cycle receives less noise, increasing the signal-to-noise ratio, there is still a notable disparity in link budget because conventional radar might be several orders of magnitude more powerful than a typical pulse-based radar.) So pulse-based UWB SAR is typically used in applications requiring average power levels in the microwatt or milliwatt range, and thus is used for scanning smaller, nearer target areas (several tens of meters), or in cases where lengthy integration (over a span of minutes) of the received signal is possible. Note, however, that this limitation is solved in chirped UWB radar systems.

The principle advantages of UWB radar are better resolution (a few millimetres using commercial off-the-shelf electronics) and more spectral information of target reflectivity.

Doppler-beam Sharpening

Doppler Beam Sharpening commonly refers to the method of processing unfocused real-beam phase history to achieve better resolution than could be achieved by processing the real beam without it. Because the real aperture of the RADAR antenna is so small (compared to the wavelength in use), the RADAR energy spreads over a wide area (usually many degrees wide in a direction orthogonal (at right angles) to the direction of the platform (aircraft). Doppler-beam sharpening takes advantage of the motion of the platform in that targets ahead of the platform return a Doppler upshifted signal (slightly higher in frequency) and targets behind the platform return a Doppler downshifted signal (slightly lower in frequency).

The amount of shift varies with the angle forward or backward from the ortho-normal direction. By knowing the speed of the platform, target signal return is placed in a specific angle "bin" that changes over time. Signals are integrated over time and thus the RADAR "beam" is synthetically reduced to a much smaller aperture – or more accurately (and based on the ability to distinguish smaller Doppler shifts) the system can have hundreds of very "tight" beams concurrently. This technique dramatically improves angular resolution; however, it is far more difficult to take advantage of this technique for range resolution.

Chirped (Pulse-compressed) Radars

A common technique for many radar systems (usually also found in SAR systems) is to "chirp" the signal. In a "chirped" radar, the pulse is allowed to be much longer. A longer pulse allows more energy to be emitted, and hence received, but usually hinders range resolution. But in a chirped radar, this longer pulse also has a frequency shift during the pulse (hence the chirp or frequency shift). When the "chirped" signal is returned, it must be correlated with the sent pulse. Classically, in analog systems, it is passed to a dispersive delay line (often a SAW device) that has the property of varying velocity of propagation based on frequency. This technique "compresses" the pulse in time – thus having the effect of a much shorter pulse (improved range resolution) while having the benefit of longer pulse length (much more signal returned). Newer systems use digital pulse correlation to find the pulse return in the signal.

Data Collection

Highly accurate data can be collected by aircraft overflying the terrain in question. In the 1980s, as a prototype for instruments to be flown on the NASA Space Shuttles, NASA operated a synthetic-aperture radar on a NASA Convair 990. However, in 1986, this plane caught fire on takeoff. In 1988, NASA rebuilt a C, L, and P-band SAR to fly on the NASA DC-8 aircraft. Called AIRSAR, it flew missions at sites around the world until 2004. Another such aircraft, the Convair 580, was flown by the Canada Centre for Remote Sensing until about 1996 when it was handed over to Environment Canada due to budgetary reasons. Most land-surveying applications are now carried out by satellite observation. Satellites such as ERS-1/2, JERS-1, Envisat ASAR, and RADARSAT-1 were launched explicitly to carry out this sort of observation.

Their capabilities differ, particularly in their support for interferometry, but all have collected tremendous amounts of valuable data. The Space Shuttle has also carried synthetic-aperture radar equipment during the SIR-A and SIR-B missions during the 1980s, as well as the Shuttle Radar Laboratory (SRL) missions in 1994 and the Shuttle Radar Topography Mission in 2000.

The Venera 15 and Venera 16 followed later by the Magellan space probe mapped the surface of Venus over several years using synthetic-aperture radar. Synthetic-aperture radar was first used by NASA on JPL's Seasat oceanographic satellite in 1978 (this mission also carried an altimeter and a scatterometer); it was later developed more extensively on the Spaceborne Imaging Radar (SIR) missions on the space shuttle in 1981, 1984 and 1994. The Cassini mission to Saturn is currently using SAR to map the surface of the planet's major moon Titan, whose surface is partly hidden from direct optical inspection by atmospheric haze.

The Mineseeker Project is designing a system for determining whether regions contain landmines based on a blimp carrying ultra-wideband synthetic-aperture radar. Initial trials show promise; the radar is able to detect even buried plastic mines. SAR has been used in radio astronomy for many years to simulate a large radio telescope by combining observations taken from multiple locations using a mobile antenna. The National Reconnaissance Office maintains a fleet of (now declassified) synthetic-aperture radar satellites commonly designated as Lacrosse or Onyx.

In February 2009, the Sentinel R1 surveillance aircraft entered service in the RAF, equipped with the SAR-based Airborne Stand-Off Radar (ASTOR) system. The German Armed Forces' (Bundeswehr) military SAR-Lupe reconnaissance satellite system has been fully operational since July 22, 2008.

Phase

Most SAR applications make use of the amplitude of the return signal, and ignore the phase data. However interferometry uses the phase of the reflected radiation. Since the outgoing wave is produced by the satellite, the phase is known, and can be compared to the phase of the return signal.

The phase of the return wave depends on the distance to the ground, since the path length to the ground and back will consist of a number of whole wavelengths plus some fraction of a wavelength. This is observable as a phase difference or phase shift in the returning wave. The total distance to the satellite (i.e. the number of whole wavelengths) is not known, but the extra fraction of a wavelength can be measured extremely accurately.

In practice, the phase is also affected by several other factors, which together make the raw phase return in any one SAR image essentially arbitrary, with no correlation from pixel to pixel. To get any useful information from the phase, some of these effects must be isolated and removed. Interferometry uses two images of the same area taken from the same position (or for topographic applications slightly different positions) and finds the difference in phase between them, producing an image known as an interferogram. This is measured in radians of phase difference and, due to the cyclic nature of phase, is recorded as repeating fringes which each represent a full 2π cycle.

Factors Affecting Phase

The most important factor affecting the phase is the interaction with the ground surface. The phase of the wave may change on reflection, depending on the properties of the material. The reflected signal back from any one pixel is the summed contribution to the phase from many smaller 'targets' in that ground area, each with different dielectric properties and distances from the satellite, meaning the returned signal is arbitrary and completely uncorrelated with that from adjacent pixels. Importantly though, it is consistent- provided nothing on the ground changes the contributions from each target should sum identically each time, and hence be removed from the interferogram.

Once the ground effects have been removed, the major signal present in the interferogram is a contribution from orbital effects. For interferometry to work, the satellites must be as close as possible to the same spatial position when the images are acquired. This means that images from two different satellite platforms with different orbits cannot be compared, and for a given satellite data from the same orbital track must be used. In practice the perpendicular distance between them, known as the *baseline*, is often known to within a few centimetres but can only be controlled on a scale of tens to hundreds of metres. This slight difference causes a regular difference in phase that changes smoothly across the interferogram and can be modelled and removed.

The slight difference in satellite position also alters the distortion caused by topography, meaning an extra phase difference is introduced by a stereoscopic effect. The longer the baseline, the smaller the topographic height needed to produce a fringe of phase change- known as the *altitude of ambiguity*. This effect can be exploited to calculate the topographic height, and used to produce a digital elevation model (DEM).

If the height of the topography is already known, the topographic phase contribution can be calculated and removed. This has traditionally been done in two ways. In the *two-pass* method, elevation data from an externally-derived DEM is used in conjunction with the orbital information to calculate the phase contribution. In the *three-pass* method two images acquired a short time apart are used to create an interferogram, which is assumed to have no deformation signal and therefore represent the topographic contribution. This interferogram is then subtracted from a third image with a longer time separation to give the residual phase due to deformation. Once the ground, orbital and topographic contributions have been removed the interferogram contains the deformation signal, along with any remaining noise. The signal measured in the interferogram represents the change in phase caused by an increase or decrease in distance from the ground pixel to the satellite, therefore only the component of the ground motion parallel to the satellite line of sight vector will cause a phase difference to be observed.

For sensors like ERS with a small incidence angle this measures vertical motion well, but is insensitive to horizontal motion perpendicular to the line of sight (approximately north-south). It also means that vertical motion and components of horizontal motion parallel to the plane of the line of sight (approximately east-west) cannot be separately resolved. One fringe of phase difference is generated by a ground motion of half the radar wavelength, since this corresponds to a whole wavelength increase in the two-way travel distance. Phase shifts are only resolvable relative to other points in the interferogram. Absolute deformation can be inferred by assuming one area in the interferogram (for example a point away from expected deformation sources) experienced no deformation, or by using a ground control (GPS or similar) to establish the absolute movement of a point.

Difficulties with InSAR

A variety of factors govern the choice of images which can be used for interferometry. The simplest is data availability- radar instruments used for interferometry commonly don't operate continuously, acquiring data only when programmed to do so. For future requirements it may be possible to request acquisition of data, but for many areas of the world archived data may be sparse. Data availability is further constrained by baseline criteria. Availability of a suitable DEM may also be a factor for two-pass InSAR; commonly 90m SRTM data may be available for many areas, but at high latitudes or in areas of poor coverage alternative datasets must be found.

A fundamental requirement of the removal of the ground signal is that the sum of phase contributions from the individual targets within the pixel remains constant between the two images and is completely removed. However there are several factors that can cause this criterion to fail. Firstly the two images must be accurately co-registered to a sub-pixel level to ensure that the same ground targets are contributing to that pixel. There is also a geometric constraint on the maximum length of the baseline- the difference in viewing angles must not cause phase to change over the width of one pixel by more than a wavelength. The effects of topography also influence the condition, and baselines need to be shorter if terrain gradients are high. Where co-registration is poor or the maximum baseline is exceeded the pixel phase will become incoherent- the phase becomes essentially random from pixel to pixel rather than varying smoothly, and the area appears noisy. This is also true for anything else that changes the contributions to the phase within each pixel, for example changes to the ground targets in each pixel caused by vegetation growth, landslides, agriculture or snow cover.

Another source of error present in most interferograms is caused by the propagation of the waves through the atmosphere. If the wave travelled through a vacuum it should theoretically be possible (subject to sufficient accuracy of timing) to use the two-way travel-time of the wave in combination with the phase to calculate the exact distance to the ground. However the velocity of the wave through the atmosphere is lower than the speed of light in a vacuum, and depends on air temperature, pressure and the partial pressure of water vapour.

It is this unknown phase delay that prevents the integer number of wavelengths being calculated. If the atmosphere was horizontally homogeneous over the length scale of an interferogram and vertically over that of the topography then the effect would simply be a constant phase difference between the two images which, since phase difference is measured relative to other points in the interferogram, would not contribute to the signal. However the atmosphere is laterally heterogeneous on length scales both larger and smaller than typical deformation signals. This spurious signal can appear completely unrelated to the surface features of the image, however in other cases the atmospheric phase delay is caused by vertical inhomogeneity at low altitudes and this may result in fringes appearing to correspond with the topography.

Producing Interferograms

The processing chain used to produce interferograms varies according to the software used and the precise application, but will usually include some combination of the following steps.

Two SAR images are required to produce an interferogram; these may be obtained pre-processed, or produced from raw data by the user prior to InSAR processing. The two images must first be co-registered, using a correlation procedure to find the offset and difference in geometry between the two amplitude images. One SAR image is then re-sampled to match the geometry of the other, meaning each pixel represents the same ground area in both images. The interferogram is then formed by cross-multiplication of each pixel in the two images, and the interferometric phase due to the curvature of the Earth is removed, a process referred to as flattening. For deformation applications a DEM can be used in conjunction with the baseline data to simulate the contribution of the topography to the interferometric phase, this can then be removed from the interferogram.

Once the basic interferogram has been produced, it is commonly filtered using an adaptive power-spectrum filter to amplify the phase signal. For most quantitative applications the consecutive fringes present in the interferogram will then have to be *unwrapped,* which involves interpolating over the 0 to 2π phase jumps to produce a continuous deformation field. At some point, before or after unwrapping, incoherent areas of the image may be masked out. The final processing stage involves geocoding the image, which resamples the interferogram from the acquisition geometry (related to direction of satellite path) into the desired geographic projection.

Software

A variety of InSAR processing packages are commonly used, several are available free or free for academic use.

- GMTSAR: An InSAR processing system based on Generic Mapping Tools- open source GNU General Public License
- IMAGINE SAR Interferometry- commercial processing package embedded in ERDAS IMAGINE remote sensing software suite, code is C++ based.
- ROI PAC- produced by NASA's Jet Propulsion Laboratory and Caltech. UNIX based, can be freely downloaded from The Open Channel Foundation.
- DORIS- processing suite from Delft University of Technology, code is C++ based, making it multi-platform portable. Distributed under GPL license from the DORIS homepage.
- Gamma Software- Commercial software suite consisting of different modules covering SAR data processing, SAR Interferometry, differential SAR Interferometry, and Interferometric Point Target Analysis, runs on Solaris, Linux, Mac OS X, Windows, large discount for Research Institutes.
- SARscape- Commercial software suite consisting of different modules covering SAR data processing, SAR and ScanSAR Interferometry, differential SAR Interferometry, Persistent Scatterers, Polarimetry and Polarimetric Interferometry, running as an extension of ArcView and ENVI under Windows, Linux and Mac OS.

- Pulsar- Commercial software suite, UNIX based.
- DIAPASON- Originally developed by the French Space Agency CNES, and maintained by Altamira Information, Commercial software suite- UNIX & Windows based
- RAT (Radar Tools)- SAR polarimetry (PolSAR), interferometry (InSAR), polarimetric interferometry (PolInSAR) and more, free software suite
- Orfeo ToolBox (OTB)- UNIX & Windows based, free software suite.

Data Sources

Early exploitation of satellite-based InSAR included use of Seasat data in the 1980s, but the potential of the technique was expanded in the 1990s, with the launch of ERS-1 (1991), JERS-1 (1992), RADARSAT-1 and ERS-2 (1995). These platforms provided the stable, well-defined orbits and short baselines necessary for InSAR. More recently, the 11-day NASA STS-99 mission in February 2000 used a SAR antenna mounted on the space shuttle to gather data for the Shuttle Radar Topography Mission. In 2002 ESA launched the ASAR instrument, designed as a successor to ERS, aboard Envisat. While the majority of InSAR to date has utilised the C-band sensors, recent missions such as the ALOS PALSAR, TerraSAR-X and COSMO SKYMED are expanding the available data in the L- and X-band.

Applications

Tectonic: InSAR can be used to measure tectonic deformation, for example ground movements due to earthquakes. It was first used for the 1992 Landers earthquake, but has since been utilised extensively for a wide variety of earthquakes all over the world. In particular the 1999 Izmit and 2003 Bam earthquakes were extensively studied. InSAR can also be used to monitor creep and strain accumulation on faults.

Volcanic: InSAR can be used in a variety of volcanic settings, including deformation associated with eruptions, inter-eruption strain caused by changes in magma distribution at depth, gravitational spreading of volcanic edifices, and volcano-tectonic deformation signals.

Early work on volcanic InSAR included studies on Mount Etna, and Kilauea, with many more volcanoes being studied as the field developed. The technique is now widely used for academic research into volcanic deformation, although its use as an operational monitoring technique for volcano observatories has been limited by issues such as orbital repeat times, lack of archived data, coherence and atmospheric errors. Recently InSAR has also been used to study rifting processes in Ethiopia.

Subsidence

InSAR, in particular subsidence caused by oil or water extraction from underground reservoirs, subsurface mining and collapse of old mines. It can also be used for monitoring the stability of built structures, and landscape features such as landslides.

DEM Generation

Interferograms can be used to produce digital elevation maps (DEMs) using the stereoscopic effect caused by slight differences in observation position between the two images. When using two images produced by the same sensor with a separation in time, it must be assumed other phase contributions (for example from deformation or atmospheric effects) are minimal. In 1995 the two ERS satellites flew in tandem with a one-day separation for this purpose. A second approach is to use two antennas mounted some distance apart on the same platform, and acquire the images at the same time, which ensures no atmospheric or deformation signals are present. This approach was followed by NASA's SRTM mission aboard the space shuttle in 2000. InSAR-derived DEMs can be used for later two-pass deformation studies, or for use in other geophysical applications.

Persistent Scatterer InSAR

Persistent or Permanent Scatterer techniques are a relatively recent development from conventional InSAR, and rely on studying pixels which remain coherent over a sequence of interferograms. In 1999, researchers at Politecnico di Milano, Italy, developed a new multi-image approach in which one searches the stack of images for objects on the ground providing consistent and stable radar reflections back to the satellite. These objects could be the size of a pixel or, more commonly, sub-pixel sized, and are present in every image in the stack.

Politecnico di Milano patented the technology in 1999 and created the spin-off company Tele-Rilevamento Europa- TRE in 2000 to commercialize the technology and perform ongoing research.

Some research centres and other companies were inspired to develop their own algorithms which would also overcome InSAR's limitations. In scientific literature, these techniques are collectively referred to as Persistent Scatterer Interferometry or PSI techniques. The term Persistent Scatterer Interferometry (PSI) was created by ESA to define the second generation of radar interferometry techniques.

Commonly such techniques are most useful in urban areas with lots of permanent structures, for example the PSI studies of European cities undertaken by the Terrafirma project. The Terrafirma project (led by Fugro NPA) provides a ground motion hazard information service, distributed throughout Europe via national geological surveys and institutions. The objective of this service is to help save lives, improve safety, and reduce economic loss through the use of state-of-the-art PSI information. Over the last 5 years this service has supplied information relating to urban subsidence and uplift, slope stability and landslides, seismic and volcanic deformation, coastlines and flood plains.

Digital Elevation Model

A digital elevation model (DEM) is a digital representation of ground surface topography or terrain. It is also widely known as a digital terrain model (DTM). A DEM can be represented as a raster (a grid of squares, also known as a heightmap when representing elevation) or as a triangular irregular network. DEMs are commonly built using remote sensing techniques, but they may also be built from land surveying. DEMs are used often in geographic information systems, and are the most common basis for digitally-produced relief maps.

Production

Mappers may prepare digital elevation models in a number of ways, but they frequently use remote sensing rather than direct survey data. One powerful technique for generating digital elevation models is interferometric synthetic aperture radar: two passes of a radar satellite (such as RADARSAT-1 or TerraSAR-X), or a single pass if the satellite is equipped with two antennas (like the SRTM instrumentation), suffice to generate a digital elevation map tens of kilometers on a side with a resolution of around ten meters.

Alternatively, other kinds of stereoscopic pairs can be employed using the digital image correlation method, where two optical images acquired with different angles taken from the same pass of an airplane or an Earth Observation Satellite (such as the HRS instrument of SPOT5 or the VNIR band of ASTER).

In 1986, the SPOT 1 satellite provided the first usable elevation data for a sizeable portion of the planet's landmass, using two-passes stereoscopic correlation. Later, further data were provided by the European Remote-Sensing Satellite (ERS) using the same method, the Shuttle Radar Topography Mission using single-pass SAR and the ASTER instrumentation on the Terra satellite using double-pass stereo pairs. Older methods of generating DEMs often involve interpolating digital contour maps that may have been produced by direct survey of the land surface; this method is still used in mountain areas, where interferometry is not always satisfactory. Note that the contour line data or any other sampled elevation datasets (by GPS or ground survey) are not DEMs, but may be considered digital terrain models. A DEM implies that elevation is available continuously at each location in the study area.

The quality of a DEM is a measure of how accurate elevation is at each pixel (absolute accuracy) and how accurately is the morphology presented (relative accuracy). Several factors play an important role for quality of DEM-derived products:

- terrain roughness;
- sampling density (elevation data collection method);
- grid resolution or pixel size;
- interpolation algorithm;
- vertical resolution;
- terrain analysis algorithm;

Methods for Obtaining Elevation Data used to Create DEMs

- Real Time Kinematic GPS
- stereo photogrammetry
- LIDAR
- Topographic maps
- Theodolite or total station
- Doppler radar
- Focus variation
- Inertial surveys.

Uses

Common uses of DEMs include:

- Extracting terrain parameters
- Modelling water flow or mass movement (for example avalanches and landslides)
- Creation of relief maps
- Rendering of 3D visualizations.
- 3d flight planning
- Creation of physical models (including raised-relief maps)
- Rectification of aerial photography or satellite imagery.
- Reduction (terrain correction) of gravity measurements (gravimetry, physical geodesy).

- Terrain analyses in geomorphology and physical geography
- Geographic Information Systems (GIS)
- Engineering and infrastructure design
- Global positioning systems (GPS)
- Line-of-sight analysis
- Base mapping
- Flight simulation
- Precision farming and forestry
- Surface analysis
- Intelligent transportation systems (ITS)
- Auto safety/Advanced Driver Assistance Systems (ADAS).

Differences between DEMs and DSMs

A *digital elevation model*- also sometimes called a *digital terrain model* (DTM)- generally refers to a representation of the Earth's surface (or subset of this), excluding features such as vegetation, buildings, bridges, etc. The DEM often comprises much of the raw dataset, which may have been acquired through techniques such as photogrammetry, LiDAR, IfSAR, land surveying, etc. A *digital surface model* (DSM) on the other hand includes buildings, vegetation, and roads, as well as natural terrain features. The DEM provides a so-called bare-earth model, devoid of landscape features. While a DSM may be useful for landscape modelling, city modelling and visualization applications, a DEM is often required for flood or drainage modelling, land-use studies, geological applications, and much more.

Sources

A free DEM of the whole world called GTOPO30 (30 arcsecond resolution, approx. 1 km) is available, but its quality is variable and in some areas it is very poor. A much higher quality DEM from the Advanced Spaceborne Thermal Emission and Reflection Radiometer (ASTER) instrument of the Terra satellite is also freely available for 99% of the globe, and represents elevation at a 30 meter resolution. A similarly high resolution was previously only available for the United States territory under the Shuttle Radar Topography Mission (SRTM) data, while most of the rest of the planet was only covered in a 3 arc-second resolution (around 90 meters). The limitation with the GTOPO30 and SRTM datasets is that they cover continental landmasses only, and SRTM does not cover the polar regions and has mountain and desert no data (void) areas. SRTM data, being derived from radar, represents the elevation of the first-reflected surface — quite often tree tops. So, the data are not necessarily representative of the ground surface, but the top of whatever is first encountered by the radar. Submarine elevation (known as bathymetry) data is generated using ship-mounted depth soundings. The SRTM30Plus dataset (used in NASA World Wind) attempts to combine GTOPO30, SRTM and bathymetric data to produce a truly global elevation model. A novel global DEM of postings lower than 12m and a height accuracy of less than 2m is expected being generated by the TanDEM-X satellite mission which started in July 2010.

The most usual grid (raster) is between 50 and 500 meters. In gravimetry e.g., the primary grid may be 50 m, but is switched to 100 or 500 meters in distances of about 5 or 10 kilometers. Many national mapping agencies produce their own DEMs, often of a higher resolution and quality, but frequently these have to be purchased, and the cost is usually prohibitive to all except public authorities and large corporations. DEMs are often a product of National LIDAR Dataset programs.

Free DEMs are also available for Mars: the MEGDR, or Mission Experiment Gridded Data Record, from the Mars Global Surveyor's Mars Orbiter Laser Altimeter (MOLA) instrument; and NASA's Mars Digital Terrain Model (DTM).

United States

The US Geological Survey produces the National Elevation Dataset, a seamless DEM for the contiguous United States, Hawaii and Puerto Rico based on 7.5' topographic mapping. As of the beginning of 2006, this replaces the earlier DEM tiled format (one DEM per USGS topographic map).

Altimeter

An altimeter is an instrument used to measure the altitude of an object above a fixed level. The measurement of altitude is called altimetry, which is related to the term bathymetry, the measurement of depth underwater.

Pressure Altimeter

Altitude can be determined based on the measurement of atmospheric pressure. The greater the altitude, the lower the pressure. When a barometer is supplied with a nonlinear calibration so as to indicate altitude, the instrument is called a pressure altimeter or barometric altimeter. A pressure altimeter is the altimeter found in most aircraft, and skydivers use wrist-mounted versions for similar purposes. Hikers and mountain climbers use wrist-mounted or hand-held altimeters, in addition to other navigational tools such as a map, magnetic compass, or GPS receiver.

Use in Hiking and Climbing

An altimeter, used along with a topographic map, can help to verify one's location. It is more reliable, and often more accurate, than a GPS receiver for measuring altitude; GPS may be unavailable, for example, when one is deep in a canyon, or may give wildly inaccurate altitudes when all available satellites are near the horizon. Because the barometric pressure changes with the weather, hikers must periodically recalibrate their altimeters when they reach a known altitude, such as a trail junction or peak marked on a topographical map.

Use in Aircraft

In it, an aneroid barometer measures the atmospheric pressure from a static port outside the aircraft. Air pressure decreases with an increase of altitude—approximately 100 hectopascals per 800 meters or one inch of mercury per 1000 feet near sea level.

The altimeter is calibrated to show the pressure directly as an altitude above mean sea level, in accordance with a mathematical model defined by the International Standard Atmosphere (ISA). Older aircraft used a simple aneroid barometer where the needle made less than one revolution around the face from zero to full scale. Modern aircraft use a "sensitive altimeter" which has a primary needle that makes multiple revolutions, and one or more secondary needles that show the number of revolutions, similar to a clock face. In other words, each needle points to a different digit of the current altitude measurement.

On a sensitive altimeter, the sea level reference pressure can be adjusted by a setting knob. The reference pressure, in inches of mercury in Canada and the US and hectopascals (previously millibars) elsewhere, is displayed in the *Kollsman window,* visible at the right side of the aircraft altimeter shown here. This is necessary, since sea level reference atmospheric pressure varies with temperature and the movement of pressure systems in the atmosphere.

In aviation terminology, the regional or local air pressure at mean sea level (MSL) is called the QNH or "altimeter setting", and the pressure which will calibrate the altimeter to show the height above ground at a given airfield is called the QFE of the field. An altimeter cannot, however, be adjusted for variations in air temperature. Differences in temperature from the ISA model will, therefore, cause errors in indicated altitude.

Sonic Altimeter

In 1931 the US Army Air Corps and General Electric tested a sonic altimeter for aircraft, which was considered more reliable and accurate than one which relied on air pressure, when heavy fog or rain was present. The new altimeter used a series of high pitched sounds like a bat to measure the distance from the aircraft to the surface, which on return to the aircraft was converted to feet shown on a gauge inside the aircraft cockpit.

Radar Altimeter

A radar altimeter, radio altimeter, low range radio altimeter (LRRA) or simply RA measures altitude above the terrain presently beneath an aircraft or spacecraft. This type of altimeter provides the distance between the plane and the ground directly below it, as opposed to a barometric altimeter which provides the distance above a pre-determined datum, usually sea level.

Principle

As the name implies, radar (radio detection and ranging) is the underpinning principle of the system. Radio waves are transmitted towards the ground and the time it takes them to be reflected back and return to the aircraft is timed. Because speed, distance and time are all related to each other, the distance from the surface providing the reflection can be calculated as the speed of the radio wave and therefore the time it takes to travel a distance are known quantities.

Alternatively, Frequency Modulated Continuous-wave radar can be used. The greater the frequency shift the further the distance travelled. This method can achieve much better accuracy than the aforementioned for the same outlay and radar altimeters that use frequency modulation are industry standard.

Invention

In 1924, American engineer Lloyd Espenschied invented the radio altimeter. However, it took 14 years before Bell Labs was able to put Espenschied's device in a form that was adaptable for aircraft use.

Civil Applications

Radar altimeters are frequently used by commercial aircraft for approach and landing, especially in low-visibility conditions and also automatic landings (autoland), allowing the autopilot to know when to begin the flare maneuver.

In civil aviation applications, radio altimeters generally only give readings up to 2,500 feet (760 m) above ground level (AGL).

Today, almost all airliners are equipped with at least one and usually several radar altimeters, as they are essential to autoland capabilities (determining height through other methods such as GPS (Global Positioning System) is not permissible under current legislation). Even older airliners from the 1960s, such as Concorde and the British Aircraft Corporation BAC 1-11 were so equipped and today even smaller airliners in the sub-50 seat class are supplied with them (such as the ATR 42 and BAe Jetstream series).

Radio altimeters are an essential part in ground proximity warning systems (GPWS), warning the pilot if the aircraft is flying too low or descending too quickly. However, radar altimeters cannot see terrain directly ahead of the aircraft, only that directly below it; such functionality requires either knowledge of position and the terrain at that position or a forward looking terrain radar which uses technology similar to a radio altimeter.

It is interesting to note that the altitude specified by the device would not match the altitude read from the standard altimeter the pilot uses. This is due to the fact that aviation is centred around True altitude, the height above Mean Sea Level (MSL), and the radio altimeter measures Absolute altitude, the height Above Ground Level (AGL). Absolute altitude is sometimes referred to as height since it is the height above the terrain directly below the aircraft, that which is provided from a radio altimeter.

Radar altimeters normally work in the E band, or Ka band or S bands for more advanced sea-level measurement. Radar altimeters also provide a reliable and accurate method of measuring height above water, when flying long sea-tracks. These are critical for use when operating to and from oil rigs.

Chapter 8

Global Navigation Satellite System

Global Navigation Satellite Systems (GNSS) is the standard generic term for satellite navigation systems ("sat nav") that provide autonomous geo-spatial positioning with global coverage. GNSS allows small electronic receivers to determine their location (longitude, latitude, and altitude) to within a few metres using time signals transmitted along a line-of-sight by radio from satellites. Receivers calculate the precise time as well as position, which can be used as a reference for scientific experiments. As of 2010, the United States NAVSTAR Global Positioning System (GPS) and the Russian GLONASS are the only two fully operational GNSS. The European Union's Galileo positioning system is a GNSS in initial deployment phase, scheduled to be operational in 2014. The People's Republic of China has indicated it will expand its regional Beidou navigation system into the global Compass navigation system by 2020. India too is developing IRNSS, which is to be operational by 2014.

The global coverage for each system is generally achieved by a constellation of 20–30 Medium Earth Orbit (MEO) satellites spread between several orbital planes. The actual systems vary, but use orbit inclinations of >50° and orbital periods of roughly twelve hours (height 20,000 km/12,500 miles).

GNSS Classification

GNSS that provide enhanced accuracy and integrity monitoring usable for civil navigation are classified as follows:

- GNSS-1 is the first generation system and is the combination of existing satellite navigation systems (GPS and GLONASS), with Satellite Based Augmentation Systems (SBAS) or Ground

Based Augmentation Systems (GBAS). In the United States, the satellite based component is the Wide Area Augmentation System (WAAS), in Europe it is the European Geostationary Navigation Overlay Service (EGNOS), and in Japan it is the Multi-Functional Satellite Augmentation System (MSAS). Ground based augmentation is provided by systems like the Local Area Augmentation System (LAAS).

- GNSS-2 is the second generation of systems that independently provides a full civilian satellite navigation system, exemplified by the European Galileo positioning system. These systems will provide the accuracy and integrity monitoring necessary for civil navigation. This system consists of L1 and L2 frequencies for civil use and L5 for system integrity. Development is also in progress to provide GPS with civil use L2 and L5 frequencies, making it a GNSS-2 system.
- Core Satellite navigation systems, currently GPS, Galileo and GLONASS.
- Global Satellite Based Augmentation Systems (SBAS) such as Omnistar and StarFire.
- Regional SBAS including WAAS (US), EGNOS (EU), MSAS (Japan) and GAGAN (India).
- Regional Satellite Navigation Systems such as China's Beidou, India's yet-to-be-operational IRNSS, and Japan's proposed QZSS.
- Continental scale Ground Based Augmentation Systems (GBAS) for example the Australian GRAS and the US Department of Transportation National Differential GPS (DGPS) service.
- Regional scale GBAS such as CORS networks.
- Local GBAS typified by a single GPS reference station operating Real Time Kinematic (RTK) corrections.

History and Theory

Early predecessors were the ground based DECCA, LORAN and Omega radio navigation systems, which used terrestrial longwave radio transmitters instead of satellites. These positioning systems broadcast a radio pulse from a known "master" location, followed by repeated pulses from a number of "slave" stations. The delay between the reception and sending of the signal at the slaves was carefully controlled, allowing the receivers to compare the delay between reception and the delay between sending. From this the distance to each of the slaves could be determined, providing a fix. The first satellite navigation system was Transit, a system deployed by the US military in the 1960s. Transit's operation was based on the Doppler effect: the satellites travelled on well-known paths and broadcast their signals on a well known frequency. The received frequency will differ slightly from the broadcast frequency because of the movement of the satellite with respect to the receiver. By monitoring this frequency shift over a short time interval, the receiver can determine its location to one side or the other of the satellite, and several such measurements combined with a precise knowledge of the satellite's orbit can fix a particular position.

Part of an orbiting satellite's broadcast included its precise orbital data. In order to ensure accuracy, the US Naval Observatory (USNO) continuously observed the precise orbits of these satellites. As a satellite's orbit deviated, the USNO would send the updated information to the satellite. Subsequent broadcasts from an updated satellite would contain the most recent accurate information about its orbit.

Modern systems are more direct. The satellite broadcasts a signal that contains orbital data (from which the position of the satellite can be calculated) and the precise time the signal was transmitted. The orbital data is transmitted in a data message that is superimposed on a code that serves as a timing reference. The satellite uses an atomic clock to maintain synchronization of all the satellites in the constellation. The receiver compares the time of broadcast encoded in the transmission with the time of reception measured by an internal clock, thereby measuring the time-of-flight to the satellite. Several such measurements can be made at the same time to different satellites, allowing a continual fix to be generated in real time.

Each distance measurement, regardless of the system being used, places the receiver on a spherical shell at the measured distance from the broadcaster. By taking several such measurements and then looking for a point where they meet, a fix is generated. However, in the case of fast-moving receivers, the position of the signal moves as signals are received from several satellites. In addition, the radio signals slow slightly as they pass through the ionosphere, and this slowing varies with the receiver's angle to the satellite, because that changes the distance through the ionosphere.

The basic computation thus attempts to find the shortest directed line tangent to four oblate spherical shells centred on four satellites. Satellite navigation receivers reduce errors by using combinations of signals from multiple satellites and multiple correlators, and then using techniques such as Kalman filtering to combine the noisy, partial, and constantly changing data into a single estimate for position, time, and velocity.

GNSS Applications

Global Navigation Satellite System (GNSS) receivers, using the GPS, GLONASS, Galileo or Beidou system, are used in many applications.

Navigation

- Automobiles can be equipped with GNSS receivers at the factory or as aftermarket equipment. Units often display moving maps and information about location, speed, direction, and nearby streets and points of interest.
- Aircraft navigation systems usually display a "moving map" and are often connected to the autopilot for en-route navigation. Cockpit-mounted GNSS receivers and glass cockpits are appearing in general aviation aircraft of all sizes, using technologies such as WAAS or LAAS to increase accuracy. Many of these systems may be certified for instrument flight rules navigation, and some can also be used for final approach and landing operations. Glider pilots use GNSS Flight Recorders to log GNSS data verifying their arrival at turn points in gliding competitions. Flight computers installed in many gliders also use GNSS to compute wind speed aloft, and glide paths to waypoints such as alternate airports or mountain passes, to aid en route decision making for cross-country soaring.
- Boats and ships can use GNSS to navigate all of the world's lakes, seas and oceans. Maritime GNSS units include functions useful on water, such as "man overboard" (MOB) functions that allow instantly marking the location where a person has fallen overboard, which simplifies rescue efforts. GNSS may be connected to the ships self-steering gear and Chartplotters using the NMEA 0183 interface. GNSS can also improve the security of shipping traffic by enabling AIS.
- Heavy Equipment can use GNSS in construction, mining and precision agriculture. The blades and buckets of construction equipment are controlled automatically in GNSS-based machine guidance systems. Agricultural equipment may use GNSS to steer automatically, or as a visual aid displayed on a screen for the driver. This is very useful for controlled traffic and row crop operations and when spraying. Harvesters with yield monitors can also use GNSS to create a yield map of the paddock being harvested.
- Bicycles often use GNSS in racing and touring. GNSS navigation allows cyclists to plot their course in advance and follow this

course, which may include quieter, narrower streets, without having to stop frequently to refer to separate maps. Some GNSS receivers are specifically adapted for cycling with special mounts and housings.

- Hikers, climbers, and even ordinary pedestrians in urban or rural environments can use GNSS to determine their position, with or without reference to separate maps. In isolated areas, the ability of GNSS to provide a precise position can greatly enhance the chances of rescue when climbers or hikers are disabled or lost (if they have a means of communication with rescue workers).
- GNSS equipment for the visually impaired is available.
- Spacecraft are now beginning to use GNSS as a navigational tool. The addition of a GNSS receiver to a spacecraft allows precise orbit determination without ground tracking. This, in turn, enables autonomous spacecraft navigation, formation flying, and autonomous rendezvous. The use of GNSS in MEO, GEO, HEO, and highly elliptical orbits is feasible only if the receiver can acquire and track the much weaker (15 - 20 dB) GNSS side-lobe signals. This design constraint, and the radiation environment found in space, prevents the use of COTS receivers. Low earth orbit satellite constellations such as the one operated by Orbcomm uses GPS receivers on all satellites.

Surveying and Mapping

- Surveying — Survey-Grade GNSS receivers can be used to position survey markers, buildings, and road construction. These units use the signal from both the L1 and L2 GPS frequencies. Even though the L2 code data are encrypted, the signal's carrier wave enables correction of some ionospheric errors. These dual-frequency GPS receivers typically cost US$10,000 or more, but can have positioning errors on the order of one centimetre or less when used in carrier phase differential GPS mode.
- Mapping and geographic information systems (GIS) — Most mapping grade GNSS receivers use the carrier wave data from only the L1 frequency, but have a precise crystal oscillator which reduces errors related to receiver clock jitter. This allows positioning errors on the order of one meter or less in real-time, with a differential GNSS signal received using a separate radio receiver. By storing the carrier phase measurements and differentially post-processing the data, positioning errors on the order of 10 centimetres are possible with these receivers.

- Several projects, including OpenStreetMap and TierraWiki, allow users to create maps collaboratively, much like a wiki, using consumer-grade GPS receivers.

- Geophysics and geology — High precision measurements of crustal strain can be made with differential GNSS by finding the relative displacement between GNSS sensors. Multiple stations situated around an actively deforming area (such as a volcano or fault zone) can be used to find strain and ground movement. These measurements can then be used to interpret the cause of the deformation, such as a dike or sill beneath the surface of an active volcano.
- Archeology — As archaeologists excavate a site, they generally make a three-dimensional map of the site, detailing where each artifact is found.
- Survey-grade GNSS receiver industry include a relatively small number of major players who specialize in the design of complex dual-frequency GNSS receivers capable of precise tracking of carrier phases for all or most of available signals in order to bring the accuracy of relative positioning down to cm-level values required by these applications. The most known companies are Javad, Leica, NovAtel, Septentrio, Topcon, Trimble.

Other Uses

- Precise time reference — Many systems that must be accurately synchronized use GNSS as a source of accurate time. GNSS can be used as a reference clock for time code generators or Network Time Protocol (NTP) time servers. Sensors (for seismology or other monitoring application), can use GNSS as a precise time source, so events may be timed accurately. Time division multiple access (TDMA) communications networks often rely on this precise timing to synchronize RF generating equipment, network equipment, and multiplexers.
- Mobile Satellite Communications — Satellite communications systems use a directional antenna (usually a "dish") pointed at a satellite. The antenna on a moving ship or train, for example, must be pointed based on its current location. Modern antenna controllers usually incorporate a GNSS receiver to provide this information.
- Emergency and Location-based services — GNSS functionality can be used by emergency services to locate cell phones. The ability to locate a mobile phone is required in the United States by E911 emergency services legislation. However, as of September 2006 such a system is not in place in all parts of the

country. GNSS is less dependent on the telecommunications network topology than radiolocation for compatible phones. Assisted GPS reduces the power requirements of the mobile phone and increases the accuracy of the location. A phone's geographic location may also be used to provide location-based services including advertising, or other location-specific information.

- Location-based games — The availability of hand-held GNSS receivers has led to games such as Geocaching, which involves using a hand-held GNSS unit to travel to a specific longitude and latitude to search for objects hidden by other geocachers. This popular activity often includes walking or hiking to natural locations. Geodashing is an outdoor sport using waypoints.
- Aircraft passengers — Most airlines allow passenger use of GNSS units on their flights, except during landing and take-off when other electronic devices are also restricted. Even though consumer GNSS receivers have a minimal risk of interference, a few airlines disallow use of hand-held receivers during flight. Other airlines integrate aircraft tracking into the seat-back television entertainment system, available to all passengers even during takeoff and landing.
- Heading information — The GNSS system can be used to determine heading information, even though it was not designed for this purpose. A "GNSS compass" uses a pair of antennas separated by about 50 cm to detect the phase difference in the carrier signal from a particular GNSS satellite. Given the positions of the satellite, the position of the antenna, and the phase difference, the orientation of the two antennas can be computed. More expensive GNSS compass systems use three antennas in a triangle to get three separate readings with respect to each satellite. A GNSS compass is not subject to magnetic declination as a magnetic compass is, and doesn't need to be reset periodically like a gyrocompass. It is, however, subject to multipath effects.
- GPS tracking systems use GNSS to determine the location of a vehicle, person, pet or freight, and to record the position at regular intervals in order to create a log of movements. The data can be stored inside the unit, or sent to a remote computer by radio or cellular modem. Some systems allow the location to be viewed in real-time on the Internet with a web-browser.
- Recent innovations in GPS tracking technology include its use for monitoring the whereabouts of convicted sex offenders, using GPS devices on their ankles as a condition of their parole. This

passive monitoring system allows law enforcement officials to review the daily movements of offenders for a cost of only $5 or $10 per day. Real time, or instant tracking is considered too costly for GPS tracking of criminals.

- GNSS Road Pricing systems charge of road users using data from GNSS sensors inside vehicles. Advocates argue that road pricing using GNSS permits a number of policies such as tolling by distance on urban roads and can be used for many other applications in parking, insurance and vehicle emissions. Critics argue that GNSS could lead to an invasion of people's privacy
- Weather Prediction Improvements — Measurement of atmospheric bending of GNSS satellite signals by specialized GNSS receivers in orbital satellites can be used to determine atmospheric conditions such as air density, temperature, moisture and electron density. Such information from a set of six micro-satellites, launched in April 2006, called the Constellation of Observing System for Meteorology, Ionosphere and Climate COSMIC has been proven to improve the accuracy of weather prediction models.
- Photographic Geocoding — Combining GNSS position data with photographs taken with a (typically digital) camera, allows one to view the photographs on a map or to lookup the locations where they were taken in a gazeteer. It's possible to automatically annotate the photographs with the location they depict by integrating a GNSS device into the camera so that co-ordinates are embedded into photographs as Exif metadata. Alternatively, the timestamps of pictures can be correlated with a GNSS track log.
- Skydiving — Most commercial drop zones use a GNSS to aid the pilot to "spot" the plane to the correct position relative to the dropzone that will allow all skydivers on the load to be able to fly their canopies back to the landing area. The "spot" takes into account the number of groups exiting the plane and the upper winds. In areas where skydiving through cloud is permitted the GNSS can be the sole visual indicator when spotting in overcast conditions, this is referred to as a "GPS Spot".
- Marketing — Some market research companies have combined GIS systems and survey based research to help companies to decide where to open new branches, and to target their advertising according to the usage patterns of roads and the socio-demographic attributes of residential zones.
- Wreck diving — A popular variant of scuba diving is known as wreck diving. In order to locate the desired shipwreck on the

bottom of the ocean floor GPS is used to navigate to the approximate location and then the shipwreck is found using an echosounder.

- Social Networking—A growing number of companies are marketing cellular phones equipped with GPS technology, offering the ability to pinpoint friends on custom created maps, along with alerts that inform the user when the party is within a programmed range. Not only do many of these phones offer social networking functions, they offer standard GPS navigation features such as audible voice commands for in-vehicle GPS navigation.

GPS Navigation Device

A GPS navigation device is any device that receives Global Positioning System (GPS) signals for the purpose of determining the device's current location on Earth. GPS devices provide latitude and longitude information, and some may also calculate altitude, although this is not considered sufficiently accurate or continuously available enough (due to the possibility of signal blockage and other factors) to rely on exclusively to pilot aircraft. GPS devices are used in military, aviation, marine and consumer product applications.

GPS devices may also have additional capabilities such as:

- containing maps, which may be displayed in human readable format via text or in a graphical format
- providing suggested directions to a human in charge of a vehicle or vessel via text or speech
- providing directions directly to an autonomous vehicle such as a robotic probe
- providing information on traffic conditions (either via historical or real time data) and suggesting alternative directions
- providing information on nearby amenities such as restaurants, fuelling stations, etc.

In other words, all GPS devices can answer the question "Where am I?", and may also be able to answer:

- which roads or paths are available to me now?
- which roads or paths should I take in order to get my desired destination?
- if some roads are usually busy at this time or are busy right now, what would be a better route to take?
- where can I get something to eat nearby or where can I get fuel for my vehicle?

Consumer Applications

Consumer GPS navigation devices include:

- Dedicated GPS navigation devices
- GPS modules that need to be connected to a computer to be used
- GPS loggers that record trip information for download. Such GPS tracking is useful for trailblazing, mapping by hikers and cyclists, and the production of geocoded photographs.
- Converged devices, including GPS Phones and GPS cameras, in which GPS is a feature rather than the main purpose of the device. Those devices may be assisted GPS or standalone (not network dependent) or both.

Automotive Navigation System

An automotive navigation system is a satellite navigation system designed for use in automobiles. It typically uses a GPS navigation device to acquire position data to locate the user on a road in the unit's map database. Using the road database, the unit can give directions to other locations along roads also in its database. Dead reckoning using distance data from sensors attached to the drivetrain, a gyroscope and an accelerometer can be used for greater reliability, as GPS signal loss and/or multipath can occur due to urban canyons or tunnels.

Some sorts can be taken out of the car and used hand-held while walking.

History

Automotive navigation systems were the subject of extensive experimentation, including some efforts to reach mass markets, prior to the availability of commercial GPS.

Most major technologies required for modern automobile navigation were already established when the microprocessor emerged in the 1970s to support their integration and enhancement by computer software. These technologies subsequently underwent extensive refinement, and a variety of system architectures had been explored by the time practical systems reached the market in the late 1980s. Among the other enhancements of the 1980s was the development of colour displays for digital maps and of CD-ROMs for digital map storage.

However, there is some question about who made the first *commercially available* automotive navigation system. There seems to be little room for doubt that Etak was first to make available a digital system that used map-matching to improve on dead reckoning instrumentation. Etak's systems, which accessed digital map information stored on standard cassette tapes, arguably made car navigation systems practical for the first time. However, Japanese efforts on both digital and analog systems predate Etak's founding.

Steven Lobbezoo developed the first commercially available satellite navigation system for cars. It was produced in Berlin from start 1984 to January 1986. Publicly presented first at the Hannover fair in 1985 in Germany, the system was shown in operation on the evening news (item in the Hannover fair) from the first German television channel in that year. It used a modified IBM PC, a large disc for map data and a flat screen, built into the glove compartment. It was called Homer (after the device from a James Bond movie).

Alpine claims to have created the first automotive navigation system in 1981. However, according to the company's own historical timeline, the company claims to have *co*-developed an analog automotive navigation product called the Electro Gyrocator, working with Honda. This engineering effort was abandoned in 1985. Although there are reports of the Electro Gyrocator being offered as a dealer option on the Honda Accord in 1981, it's not clear whether an actual product was released, whether any customers took delivery of an Electro Gyrocator-equipped Accord, or even whether the unit appeared in any dealer showrooms; Honda's own official history appears to pronounce the Electro Gyrocator as not practical.

Honda claims to have created the first navigation system starting in 1983, and culminating with general availability in the 1990 Acura Legend. The original analog Electro Gyrocator system used an accelerometer to navigate using inertial navigation, as the GPS system was not yet generally available. However, it appears from Honda's concessions in their own account of the Electro Gyrocator project that Etak actually trumped Honda's analog effort with a truly practical digital system, albeit one whose effective range of operation was limited by the availability of appropriately digitized street map data.

[...] progress in digital technology would not stop simply because Honda had turned its attention to analog. In 1985, for example, the U.S. company ETAK introduced its own digital map navigation system. Although the system's effective range-the area of geographical coverage-was limited, the announcement was a dour one for Nakamura and his staff. Therefore, ultimately the development of a practical analog system was shelved. The staff experienced indescribable feelings of disappointment. The development of [Honda's] digital map navigation system resumed in 1987, following a three-year hiatus.

Both Mitsubishi Electric and Pioneer claim to be the first with a GPS-based auto navigation system, in 1990. Also in 1990, a draft patent application was filed within Digital Equipment Co. Ltd. for a multifunction device called PageLink that had real-time maps for use in a car listed as one of its functions. Magellan, a GPS navigation system manufacturer, claims to have created the first GPS-based vehicle navigation system in the U.S. in 1995.

In 1995, Oldsmobile introduced the first GPS navigation system available in a production car, called GuideStar. There also was an Oldsmobile navigation system available as an option as early as 1994 called the Oldsmobile Navigation/Information System. It was an option on the Oldsmobile Eighty Eight. However it was not until 2000 that the United States made a more accurate GPS signal available for civilian use.

Technology

Visualization: Navigation systems may (or may not) use a combination of any of the following:

- top view for the map
- top view for the map with the map rotating with the automobile (so that "up" on the map always corresponds to "forward" in the vehicle)
- bird's-eye view for the map or the next curve
- linear gauge for distance, which is redundant if a rotating map is used
- numbers for distance
- schematic pictograms
- voice prompts.

Road Database

Contents: The road database is a vector map of some area of interest. Street names or numbers and house numbers are encoded as geographic coordinates so that the user can find some desired destination by street address.

Points of interest (waypoints) will also be stored with their geographic coordinates. Point of interest specialties include speed cameras, fuel stations, public parking, and "parked here" (or "you parked here").

Contents can be produced by the user base as their cars drive along existing streets (Wi-Fi) and communicating via the internet, yielding a free and up-to-date map.

Map Formats

Formats are almost uniformly proprietary; there is no industry standard for satellite navigation maps, although NAVTEQ are currently trying to address this with S-Dal. The map data vendors such as Tele

Atlas and NAVTEQ create the base map in a standard format GDF, but each electronics manufacturer compiles it in an optimized, usually proprietary format. GDF is not a CD standard for car navigation systems. GDF is used and converted onto the CD-ROM in the internal format of the navigation system.

CARiN

CARiN Database Format (CDF) is a proprietary navigation map format created by Philips Car Systems (this branch was sold to Mannesman VDO, VDO/Dayton in 1998, to Siemens VDO in 2002, and Continental in 2007.) and is used in a number of navigation-equipped vehicles. The 'CARiN' portmanteau is derived from Car Information and Navigation.

S-Dal

This is a proprietary map format published by NAVTEQ, who released it royalty free in the hope that it would become an industry standard for digital navigation maps. Vendors currently using this format include:

- Microsoft
- Magellan
- Pioneer
- Panasonic
- Clarion
- InfoGation.

The format has not been very widely adopted by the industry.

Physical Storage Format

The Physical Storage Format (PSF) initiative is an industry grouping of car manufacturers, navigation system suppliers and map data suppliers whose objective is the standardization of the data format used in car navigation systems, as well as allow a map update capability. Standardization would improve interoperability, specifically by allowing the same navigation maps to be used in navigation systems from 19 manufacturers. Companies involved include BMW, Volkswagen, Daimler, Renault, ADIT, Aisin AW, Alpine Electronics, Navigon, Bosch, DENSO, Mitsubishi, Harman Becker, Panasonic, PTV, Continental AG, Clarion, NAVTEQ, Tele Atlas and Zenrin.

Media

The road database may be stored in solid state read-only memory (ROM), optical media (CD or DVD), solid state flash memory, magnetic media (hard disk), or a combination. A common scheme is to have a base map permanently stored in ROM that can be augmented with detailed information for a region the user is interested in. A ROM is always programmed at the factory; the other media may be preprogrammed, downloaded from a CD or DVD via a computer or wireless connection (bluetooth, Wi-Fi), or directly used utilizing a card reader.

Some navigation device makers provide free map updates for their customers. These updates are often obtained from the vendor's website, which is accessed by connecting the navigation device to a PC.

Real-time Data

Some newer systems can not only give precise driving directions, they can also receive and display information on traffic congestion and suggest alternate routes. These may use either TMC, which delivers coded traffic information using radio RDS, or by GPRS/3G data transmission via mobile phones.

One key type of real-time data is traffic information, which includes:

- Real-time data about free/full parkings;
- Nearest public transport lines and prices, to go to a destination, when there is a jam.

Other real-time data includes weather broadcasting, etc.

Integration and other Functions

- The colour LCD screens on some automotive navigation systems can also be used to display television broadcasts or DVD movies.
- A few systems integrate (or communicate) with mobile phones for hands-free talking and SMS messaging (i.e., using Bluetooth or Wi-Fi).
- Automotive navigation systems can include personal information management for meetings, which can be combined with a traffic and public transport information system.

Controversy

Safety Features: Vehicles produced by Subaru and Lexus, as well as Lexus' parent company, Toyota, lock out many of the features when the vehicle is in motion. The manufacturers claim this is a safety feature to avoid the driver being distracted. Many users have complained that passengers are not able to enter destinations while in motion, even though it is safe to do so. Additionally, drivers have complained that it is often more dangerous to pull off a highway and stop than it would be to enter a destination into the system.

Misdirection

A number of road accidents in the UK have been attributed to misdirection by satellite navigation systems. On May 11, 2007, a driver followed satellite navigation instructions in the dark and her car was hit by a train on a rail crossing that was not shown on the system. In Exton, Hampshire, the County Council erected a sign warning drivers to ignore their "sat nav" system and to take another route, because the street was too narrow for vehicular traffic and property damage resulted from vehicles getting stuck. On March 25, 2009, a man drove down a steep mountain path and almost off of a cliff after he was allegedly directed by his portable GPS system. He was finally stopped by a wire fence. Misdirection can also occur when a road is altered either permanently or temporarily, such as during road re-construction.

GPS vs Speed Camera Accuracy

In July 2007, an Australian man successfully overturned a speeding conviction after evidence from a GPS navigational track proved that he did not exceed the speed limit.

Other Functions

- Golf Carts may have integrated GPS rangefinders tailored to specific golf courses, providing interactive course maps and live readings of distance measurements to the green.
- Many systems can give information on nearby points of interest (POIs), such as restaurants, cash machines and gas stations. Some navigation devices use this feature to store the location of known speed traps or speed cameras, and can alert the driver in much the same way as a radar detector. GPS may also be integrated into actual radar detection devices to enhance accuracy, and in some cases, implement a logic system where the system only alerts if the driver is travelling above the speed limit or in the direction to be 'caught.' Unlike radar detectors, GPS-based speed trap warnings are currently legal in many countries.
- Some systems feature internet connectivity, either via Bluetooth to a mobile phone (in which case the device can typically also be used for hands-free calling), or with a built in GSM SIM card. This connectivity can be used for up-to-date traffic information, to find fuel prices, as well as to search for local distances. Such devices include the TomTom LIVE series, and the Garmin nuvi 1690.
- The radio dispatching of taxicabs have been phased out in several countries in favor of GPS technology plus some form of mobile networking with on board computers. The central

dispatch computer keeps track of all vehicles in its fleet, and automatically selects the nearest cab to respond to a passenger request.

- Advanced car security vehicle tracking systems can relay the vehicle's location via cellular phone services in case of loss or theft. The technology can also be used to manage fleet vehicles, in which case it's known as automatic vehicle location.
- A very basic form of GPS navigation is used on public buses in Taipei, where the location and sequence of bus stops for a particular route are programmed. The computer announces the approaching and upcoming bus stops and repeats the information on a dot-matrix display, all without intervention from the driver. This service was once provided based on tire revolutions and odometer mileage, which is not nearly as reliable as a GPS enabled system.

Retrofitting of GPS

A vehicle can be retrofitted with a GPS navigation device unit if it did not originally have one. There are three approaches that can be taken here:

Portable GPS

This type of GPS navigation device is not permanently integrated into the vehicle, having only a simple bracket to mount the device on the surface of the dashboard and powered via the car cigarette lighter. This class of GPS unit does not require professional installation and can typically be used as handheld device, too. Benefits of this type of GPS unit include low cost as well as the ability to move them easily to other vehicles. Their portability means they are easily stolen if left inside the vehicle. Furthermore, not having a compass, accelerometer or inputs from the vehicle's speed sensors, means that they cannot navigate as accurately by dead reckoning as some built-in devices when there's no GPS signal. More modern portable devices such as the TomTom 920, have an inbuilt accelerometer to try to address this.

A portable automotive navigation system kit generally includes:

- Mini-USB sync cable
- AC adaptor
- Car charger
- Car mount kit
- Pouch
- Wrist band

- External antenna (optional by model)
- Stylus
- Battery pack
- Document kit
- SD card with preload map (sometimes capable of shuffling MP3 playlists)
- Companion CD-ROM
- Navigation software CD-ROM.

Many vehicle manufacturers offer a GPS navigation device as an option in their vehicles. Customers whose vehicles did not ship with GPS can therefore purchase and retrofit the original factory-supplied GPS unit. In some cases this can be a straightforward "plug-and-play" installation if the required wiring harness is already present in the vehicle. However, with some manufacturers, new wiring is required, making the installation more complex.

The primary benefit of this approach an integrated and factory-standard installation. Many original systems also contain a gyrocompass or accelerometer and may accept input from the vehicle's speed sensors, thereby allowing them to navigate via dead reckoning when a GPS signal is temporarily unavailable. However, the costs can be considerably higher than other options. In some cases, it may even be more economical to buy a similar vehicle that already has a factory-fitted GPS.

Aftermarket

A number of manufacturers supply aftermarket GPS navigation devices that can be integrated permanently into the vehicle. A typical location for such an installation is the DIN slot for the radio/tape/CD. However, in extreme cases, the dashboard may also be remodeled to accommodate the unit. This approach can be considered a tradeoff between the previous two options. Benefits include a more secure and better cosmetic finish than a portable device, and lower cost compared to the installation of an original factory-supplied GPS.

Alternatives

Smartphones with GPS, and other navigation devices, may also be used without installing in a car.

SMS

Establishing points of interest in real-time and transmitting them via GSM cellular telephone networks using the Short Message Service (SMS) is referred to as Gps2sms. Some vehicles and vessels are equipped with hardware that is able to automatically send an SMS text message when a particular event happens, such as theft, anchor drift or breakdown. The receiving party (e.g., a tow truck) can store the waypoint in a computer system, draw a map indicating the location, or see it in an automotive navigation system.

Personal Navigation Assistant

A Personal Navigation Assistant (PNA) also known as Personal Navigation Device or Portable Navigation Device (PND) is a portable electronic product which combines a positioning capability (such as GPS) and navigation functions. Some PNA devices are PDAs with limited features and can be unlocked.

History

The earliest PNAs were hand-held GPS units (circa mid 1980s) which were capable of displaying the user's location on an electronic map. These units included simple navigation functions such as course-to-steer and course-made-good. This first generation of PNAs were used by the US military.

Market Developments

According to the analyst firm Berg Insight, there were more than 150 million turn-by-turn navigation systems worldwide in mid 2009, including about 35 million factory installed and aftermarket in-dash navigation systems, over 90 million Personal Navigation Devices (PNDs) and an estimated 28 million navigation-enabled mobile handsets with GPS.

The term PNA has come into widespread use with the growing popularity of automobile navigation systems. The latest generation of PNA have sophisticated navigation functions and feature a variety of user interfaces including maps, turn-by-turn guidance and voice instructions. To reduce total cost of ownership and time to market, most modern PNA devices such as those made by Garmin Ltd., Mio Technology Ltd. or TomTom International BV. are running an off-the-shelf embedded operating system such as Windows CE or Embedded Linux on commodity hardware with OEM versions of popular PDA Navigation software packages such as TomTom Navigator, I-GO 2006, Netropa IntelliNav iGuidance, or Destinator. Other manufacturers like Garmin and Magellan prefer to bundle their own software developed in house. Because many of these devices use an embedded OS, many technically inclined users find it easy to modify PNAs to run third party software and use them for things other than navigation, such as a low-cost audio-video player or PDA replacement.

Meantime, Nokia, Samsung Electronics, Motorola and other handset makers will sell 162 million GPS phones in 2007, dwarfing the 20 million units Garmin and TomTom have forecast they will sell combined, according to iSuppli, a leading market researcher in California.

The inclusion of Google Maps Navigation in the iPhone and Motorola Droid, and Nokia's announcement of free Ovi Maps, suggest that PNA will be rapidly integrated into mobile phones.

Mobile Phones with GPS Capability

Due in part to regulations encouraging mobile phone tracking, including E911, the majority of GPS receivers are built into mobile telephones, with varying degrees of coverage and user accessibility. Commercial navigation software is available for most 21st century smartphones as well as some Java-enabled phones that allows them to use an internal or external GPS receiver (in the latter case, connecting via serial or Bluetooth). Some phones with GPS capability work by assisted GPS (A-GPS) only, and do not function when out of range of their carrier's cell towers. Others can navigate worldwide with satellite GPS signals as a dedicated portable GPS receiver does, upgrading their operation to A-GPS mode when in range. Still others have a hybrid positioning system that can use other signals when GPS signals are inadequate.

More bespoke solutions also exist for smartphones with inbuilt GPS capabilities. Some such phones can use tethering to double as a wireless modem for a laptop, while allowing GPS-navigation/localisation as well. One such example is marketed by Verizon Wireless in the United States, and is called VZ Navigator. The system uses gpsOne technology to determine the location, and then uses the mobile phone's data connection to download maps and calculate navigational routes. Other products including iPhone are used to provide similar services. Nokia gives Ovi Maps free on its smartphones and maps can be preloaded. According to market research from the independent analyst firm Berg Insight, the sales of GPS-enabled GSM/WCDMA handsets was 150 million units in 2009. while only 40 million separate GPS receivers were sold.

GPS navigation applications for mobile phones include Waze.

Laptop PC GPS

Various software companies have made available GPS road navigation software programs for in-vehicle use on laptop computers. Benefits of GPS on a laptop include larger map overview, ability to use the keyboard to control GPS functions, and some GPS software for laptops offers advanced trip-planning features not available on other platforms. Laptop computers allow for other uses beside GPS.

GPS Modules

Other GPS devices need to be connected to a computer in order to work. This computer can be a home computer, laptop or even a PDAs, or smartphones. Depending on the type of computer and available connectors, connections can be made through a serial or USB cable, as well as Bluetooth, CompactFlash, SD, PCMCIA and the newer ExpressCard. Some PCMCIA/ExpressCard GPS units also include a wireless modem. Devices usually do not come with preinstalled GPS navigation software, thus once purchased the user must install or write their own navigation software. As the user can choose which navigation software to use, it can be better matched to their personal taste. It is very common for a PC-based GPS receiver to come bundled with a navigation software suite. Also, GPS modules are significantly cheaper than complete stand-alone systems (around 50-100 €). The software may include maps only for a particular region, or the entire world (if software such as Google Maps, Networks in Motion's AtlasBook mobile navigation platform, etc. is used).

Examples of Bluetooth GPS devices are:

- Holux GPSlim236.

For examples of USB GPS devices, see:

- Globalsat BU-303 GPS
- Holux ?
- DeLorme Earthmate LT-40 with Street Atlas USA 2009
- Haicom HI-204 III USB GPS
- Canmore GT-730F USB GPS
- Navman GPS e Series
- PlayStation Portable PSP-290.

Examples of CF GPS devices are:

- Globalsat BC-337 SiRF Star III Compact Flash)
- Gophers SiRF Atlas V 500MHz Compact Flash
- Holux GR-271 Slim Compact Flash
- Haicom Hi-303III CompactFlash.

Examples of ExpressCard GPS devices with embedded modem are:

- Sony Ericsson ec400g

Some hobbyists have also made some GPS devices and open-sourced the plans. An example is the Elektor GPS units. These are based around a SirFStar 3 chip and are comparable to their commercial counterparts.

Commercial Aviation

Commercial aviation applications include GPS devices that calculate location and feed that information to large multi-input navigational computers for autopilot, course information and correction displays to the pilots, and course tracking and recording devices.

Restrictions on Civilian Use

The U.S. Government controls the export of some civilian receivers. All GPS receivers capable of functioning above 18 kilometers (11 mi) altitude and 515 metres per second (1,001 kn) are classified as munitions (weapons) that U.S. State Department export licenses are required. These limits attempt to prevent use of a receiver in a ballistic missile. They would not prevent use in a cruise missile because their altitudes and speeds are similar to those of ordinary aircraft.

This rule applies even to otherwise purely civilian units that only receive the L1 frequency and the C/A (Clear/Acquisition) code and cannot correct for Selective Availability (SA), etc.

Disabling operation above these limits exempts the receiver from classification as a munition. Vendor interpretations differ. The rule targets operation given the combination of altitude and speed, while some receivers stop operating even when stationary. This has caused problems with some amateur radio balloon launches that regularly reach 30 kilometers (19 mi).

Military

As of 2009, military applications of GPS include:

- Navigation: GPS allows soldiers to find objectives, even in the dark or in unfamiliar territory, and to coordinate troop and supply movement. In the US armed forces, commanders use the *Commanders Digital Assistant* and lower ranks use the *Soldier Digital Assistant*.
- Target tracking: Various military weapons systems use GPS to track potential ground and air targets before flagging them as hostile. These weapon systems pass target coordinates to precision-guided munitions to allow them to engage targets accurately. Military aircraft, particularly in air-to-ground roles, use GPS to find targets (for example, gun camera video from AH-1 Cobras in Iraq show GPS co-ordinates that can be viewed with special software.
- Missile and projectile guidance: GPS allows accurate targeting of various military weapons including ICBMs, cruise missiles

and precision-guided munitions. Artillery projectiles. Embedded GPS receivers able to withstand accelerations of 12,000 *g* or about 118 km/s^2 have been developed for use in 155 millimetres (6.1 in) howitzers.

- Search and Rescue: Downed pilots can be located faster if their position is known.
- Reconnaissance: Patrol movement can be managed more closely.
- GPS satellites carry a set of nuclear detonation detectors consisting of an optical sensor (Y-sensor), an X-ray sensor, a dosimeter, and an electromagnetic pulse (EMP) sensor (W-sensor), that form a major portion of the United States Nuclear Detonation Detection System.

Awards

Two GPS developers received the National Academy of Engineering Charles Stark Draper Prize for 2003:

- Ivan Getting, emeritus president of The Aerospace Corporation and engineer at the Massachusetts Institute of Technology, established the basis for GPS, improving on the World War II land-based radio system called LORAN (*Lo*ng-range *R*adio *A*id to *N*avigation).
- Bradford Parkinson, professor of aeronautics and astronautics at Stanford University, conceived the present satellite-based system in the early 1960s and developed it in conjunction with the U.S. Air Force. Parkinson served twenty-one years in the Air Force, from 1957 to 1978, and retired with the rank of colonel.

GPS developer Roger L. Easton received the National Medal of Technology on February 13, 2006.

On February 10, 1993, the National Aeronautic Association selected the GPS Team as winners of the 1992 Robert J. Collier Trophy, the nation's most prestigious aviation award. This team combines researchers from the Naval Research Laboratory, the U.S. Air Force, the Aerospace Corporation, Rockwell International Corporation, and IBM Federal Systems Company. The citation honours them "for the most significant development for safe and efficient navigation and surveillance of air and spacecraft because the introduction of radio navigation 50 years ago."

Bibliography

Ahrens, Donald C.: *Meteorology Today: An Introduction to Weather, Climate, and the Environment*, West Publishing Company, St. Paul, Minnesota, 1991.

Allan, T.D.: *Satellite Microwave Remote Sensing*, John Wiley and Sons, New York, 1983.

Bankert, R.L. and P.M. Tag: *Automating the Subjective Analysis of Satellite-Observed Tropical Cyclone Intensity. Proceedings*, Conference on Hurricanes and Tropical Meteorology, American Meteorological Society, Boston, MA, 1997.

Bhatta, B : *Remote Sensing and GIS,* Oxford University Press, Delhi, 2008.

Carleton, Andrew M. : *Satellite Remote Sensing in Climatology*, London, CRC Press, 1991.

Curtis, L. F. : *Introduction to Environmental Remote Sensing*, London, New York, 1992.

Drager, Dwight L. and Thomas R. Lyons: *Remote Sensing in Cultural Resource Management: The San Juan Basin Project,* National Park Service, Washington, DC, 1983.

Dunai, T.J.: *Cosmogenic Nucleides,* Cambridge University Press, U.K., 2010.

Egan, Walter G. : *Photometry and Polarization in Remote Sensing*, New York, 1985.

Frances Joan : *Cultural Resources Remote Sensing,* Washington DC, National Park Service, Cultural Resources Management Division, 1980.

Golledge, R. and R. Stimson: *Spatial Behavior: A Geographic Perspective.* New York: The Guilford Press, 1997.

Hartshorne, Richard : *Perspective on the Nature of Geography*, Chicago, Rand McNally, 1959.

Inkpen, Robert: *Science, Philosophy and Physical Geography*, Routledge, London.

Johnson, Jay K.: *Remote Sensing in Archaeology*, University of Alabama Press, Tuscaloosa, 2006.

Juppenlatz, Morris: *Geographic Information Systems and Remote Sensing*, McGraw Hill, New York, 1996.

Kumar Yadav, Surendra : *Soil Ecology : Remote Sensing and Geographic Information System in Management,* APH, Delhi, 2007.

Langran, G.: *Time in Geographic Information Systems.* Bristol: Taylor & Francis, 1992.

Lillesand, T. M. and Kiefer, R. W. : *Remote Sensing and Image Interpretation*, New York, 1987.

Mack, Pamela Etter : *Viewing the Earth: The Social Construction of the Landsat Satellite System*, Cambridge, MIT Press, 1900.

May, D.A., J. Sandidge, R. Holyer, and J.D. Hawkins: *SSM/I Derived Tropical Cyclone Intensities*, American Meteorological Society, Boston, MA, 1997.

Miller, H. and S.-L. Shaw: *Geographic Information Systems for Transportation: Principals and Applications.* New York: Oxford, 2001 University Press.

Mulder, M.A.: *Application of Remote Sensing to Soils*, Developments in Soil Science, Elsevier, New York. 1987.

Peuquet, D.: *Representations of Space and Time.* New York: Guilford Press, 2002.

Przyrbyla, J.: *Practical Considerations for GIS.* Redlands, California: 2010 GeoDesign Summit, 06-08 January 2010.

Raymond M. : *Laser Remote Sensing: Fundamentals and Applications*, Malabar, Fla., Krieger Pub. 1992.

Ritter, D.F., Kochel, R.C., and Miller, J.R.: *Process Geomorphology*, Wm.C. Brown Publishers, Dubuque, Iowa, 1995.

Schanda, Erwin : *Physical Fundamentals of Remote Sensing,* Berlin, New York, 1986.

Scheidegger, Adrian E.: *Morphotectonics*, Springer, Berlin, 2004.

Smithson, Peter: *Fundamentals of the Physical Environment*, Routledge, London, 2002.

Strahler, Alan; Strahler Arthur: *Introducing Physical Geography*, Wiley, New York, 2006.

Turcotte, D.L.; Schubert, G.: "Plate Tectonics" *Geodynamics,* Cambridge University Press, Cambridge, 2002.

Viles, Heather: *Biogeomorphology*, Basil Blackwell, Oxford, 1988.

Wainwright, John; Mulligan, M.: *Environmental Modelling: Finding Simplicity in Complexity*, John Wiley and Sons Ltd, London, 2003.

Williams, Paul W.: *Catena Supplement 25,* Catena Verlag, Cremlingen-Destedt, Germany, 1993.

Index

❑❑❑